Taking Biology Seriously

Taking Biology Seriously

What Biology Can and Cannot Tell Us About Moral and Public Policy Issues

Inmaculada de Melo-Martín

ROWMAN & LITTLEFIELD PUBLISHERS, INC.
Lanham • Boulder • New York • Toronto • Oxford

ROWMAN & LITTLEFIELD PUBLISHERS, INC.

Published in the United States of America
by Rowman & Littlefield Publishers, Inc.
A wholly owned subsidiary of The Rowman & Littlefield Publishing Group, Inc.
4501 Forbes Boulevard, Suite 200, Lanham, Maryland 20706
www.rowmanlittlefield.com

PO Box 317
Oxford
OX2 9RU, UK

British Library Cataloguing in Publication Information Available

Library of Congress Cataloging-in-Publication Data

Melo-Martín, Inmaculada de.
 Taking biology seriously : what biology can and cannot tell us about moral and
 public policy issues / Inmaculada de Melo-Martín.
   p. cm.
 Includes bibliographical references and index.
 ISBN 0-7425-4920-8 (cloth : alk. paper) — ISBN 0-7425-4921-6 (pbk. : alk. paper)
 1. Human genetics—Social aspects. 2. Human biology—Social aspects.
 3. Cloning—Social aspects. I. Title.

 QH438.7.M45 2005
 174.2—dc22   2005013430

Printed in the United States of America

⊗™ The paper used in this publication meets the minimum requirements of American
National Standard for Information Sciences—Permanence of Paper for Printed Library
Materials, ANSI/NISO Z39.48-1992.

For Craig, Damián, and Martín, the other men of my family

Additional Praise for
Taking Biology Seriously

"Professor de Melo-Martín shows herself to be as knowledgeable as she says philosophers and others making ethical and political claims should be. Though many will disagree with her views, it is welcome to have a clearly stated and well-argued position with which others can choose to disagree or not."

—**Paul Durbin**, University of Delaware

"Dr. de Melo-Martín offers an insightful and sensible account of the mistakes that philosophers, scientists, and policy makers commit when drawing conclusions about genetics."

—**Gonzalo Munevar**, Lawrence Technical College

Contents

Acknowledgments

As with most works, this book would not have been possible without the assistance, guidance, support, and care of many people. Some have been involved directly in the development of this project through advice, criticisms, and suggestions; others have offered needed love and support; and still others have done both.

For their support and encouragement through the whole process of thinking and writing about the present topics, I am especially grateful to Gonalo Munévar and Craig Hanks. They read and commented on several earlier drafts without complaint, and their suggestions have, without any doubt, made this work better than it could have been otherwise.

I have received insightful and generous comments from several friends and colleagues. Thanks to Sharyn Clough for her support and suggestions, and to Evelyn Brister for using her vacation time to read an earlier draft of this manuscript. I am also grateful to Kristen Intemann and Anita Ho, who, in spite of the craziness of their lives at that moment, took the time to carefully read this work and offer detailed comments and suggestions. Thank you also to Miguel Martinez-Saenz for the interminable phone conversations and for reading the book and offering useful comments, and to a number of anonymous reviewers for their advice and helpful criticisms.

I am also grateful to my chair, Conrad Kaczkowski. He always gives me great schedules and does as much as he can to support my research efforts. Thank you also to St. Mary's University for providing me with funds for a sabbatical year and for graduate work on an MS in biology, both of which have been extremely helpful for the work on this book.

My family, in spite of the distance, is a continuous source of care, love, and support. Chus is always there for me. My friends Bonnie Walker, David Miron, and Chip Hughes always make sure I get time to go out and enjoy good food and wine.

My greatest thanks are to Craig Hanks, for his patience, love, inspiration, and encouragement. He makes my life so much easier and more joyful and fulfilling than it otherwise would be.

Sections of this book's chapters have appeared as articles in journals. These include "Biological Explanations and Social Responsibility," *Studies in History and Philosophy of Biological and Biomedical Sciences* 34, no. 2 (2003): 345–58; and "On Cloning Human Beings," *Bioethics* 16, no. 3 (2002): 246–65. I thank the publishers of these journals for permission to use these materials.

Chapter One

Misunderstanding Biology: Epistemological, Scientific, and Moral Problems

Is it immoral to clone a human being? Do we have an obligation not to re-produce when doing so would bring a gravely ill child into the world? Should we experiment on human embryos? Are the social responsibilities of communities diminished when particular biological traits and behaviors are genetically determined? Should we implement genetic testing programs for complex diseases? Do we have an obligation to select the best child out of the possible children we could have? Is it moral to use genetic enhancements? Do we have a duty to obtain information about our genetic endowments? Do we have an obligation to share information about our genetic makeup with our family members? One need not be an expert to be aware of the ways in which our current biomedical knowledge and our technological abilities are presenting us with difficult moral and political decisions.

All of these questions have in common the intermingling of biomedical science and technology with moral responsibilities or policy concerns. It is only appropriate, then, that we evaluate the relationships between biomedical knowledge and its consequences for morality and public policy. But why should biological knowledge be of concern to ethicists and public policy makers? It is obvious that if one wants to make claims about what follows ethically or politically from particular biomedical advances, one must be familiar with biomedical science and technology. For example, we must know something about how genes work in order to wonder whether we have an obligation to obtain genetic information about ourselves or whether justice requires that we allow people to enhance genetic traits. Or, we must know something about the biology of cancer to promote health policies that advise people not to smoke. Without this particular scientific and technological knowledge, moral and public policy questions about these biomedical advances would evidently not

1

arise. In other occasions, moral and public policy questions related to scientific knowledge arise, but a lack of information prevents us from giving definite answers. For instance, many of the debates about the safety of transgenic organisms are the result of inconclusive evidence. Attaining the right information on this issue would allow us to offer answers to moral and public policy issues, such as what kind of regulatory mechanisms we should have in place. Furthermore, in many cases, biological knowledge presumably grounds the answers to particular ethical and public policy questions. For example, a defense of the obligation to select the best child out of the possible children we could have is grounded in part on the belief that traits such us intelligence, memory, beauty, or musical ability are traits mainly controlled by genes. Similarly, the assumption that our genome is somehow related to our individuality grounds arguments about the immorality of cloning human beings. Hence, biological knowledge does, and should, inform many of the answers we give when dealing with ethical and public policy issues related to biomedical technologies. Of course this does not mean that biology alone can offer responses to these problems. Careful evaluation and defense of ethical principles, social context, setting of priorities, and economic possibilities are also necessary. Consequently, to require that biology inform our discussions on moral and political issues related to biomedical science and technology only means that biology should not be ignored. It does not mean that biological knowledge can be used as a trump card.

Paradoxically, both critics and supporters of the certainly desirable incorporation of biology and genetics into our discussions of moral behavior and public policy have tended to present biology's consequences for morality and politics in flawed ways. Critics have tended to see human biology as an inexorable force that, unless tightly controlled, would threaten to destroy many hard-fought social, political, and ethical advances. Ironically, supporters of incorporating biology into the social sciences and the humanities seem to have a similar conception of human biological nature. They, however, have opted for embracing human biology as the way to solve the problems that plague us and have argued that biological facts ought to be separated from human values. Presumably, by doing so we would be able to avoid what appear to be the undesirable consequences of conceding importance to human biology.

Both critics and supporters misconceive the role of biology. Human biology is neither a frightful nor a fantastic inescapable force whose consequences can be evaluated independently of the social, political, and ethical context in which human beings live. It is humans' biology to be social creatures, to construct societies, to develop legal and social institutions, and to follow moral codes. Paraphrasing Dobzhansky's famous claim, nothing in human biology can make sense except in the light of humans' social environment. Any evaluation of human biology that ignores this point misunderstands biology and its consequences for philosophy and social policy. Any such evaluation does not take biology seriously.

The purpose of this book is to evaluate some of the misunderstandings present in many discussions of biology and its consequences for morality and public policy. The mistakes I will analyze here are the following: First, those concerned with the implications of human biology for morality and public policy can commit epistemological errors. In these cases, we are told mistakenly that knowledge of a particular aspect of our biology will give us knowledge about some other aspects of human life. Second, they can misunderstand the scientific issues at stake. We are presented, then, with moral and public policy consequences that follow from a misreading of current biological knowledge. Third, they might have an erroneous conception of morality. Here, although they may have the science right, and they may have the consequences of the science right, the problem is that those consequences may not have the moral import attributed to them.

An evaluation of these mistakes will help us understand correctly what current biological knowledge can tell us about ourselves as biological beings. It will also give us information about what follows for our moral conduct and public policies from a correct understanding of both human biology and morality. We can expect that it will assist us in evaluating the adequacy or inadequacy of present public policies that are presumably grounded on biological knowledge. Finally, it can guide us toward wiser decisions when dealing with the increasing dilemmas we face when making use of our scientific and technological abilities.

OVERVIEW

Not just philosophers, but also scientists, those in the medical profession, social scientists, policy makers, and the public at large have been quick to embrace the successes of the genetic revolution. The sequencing of the human genome, for example, is publicized as the new approach to solve the medical problems that afflict humanity. Discussions about whether genetic enhancement of human traits is right or wrong, presentations of genetic therapy as the new revolution in medicine, questions about cloning human beings, and concerns about enacting public policies that would prevent the use of genetic information to discriminate against people all seem to presuppose that the new science of genetics has the power to drastically transform people's lives and the world.[1]

It is not a goal of this project to offer an exhaustive evaluation of all the cases in which biomedical science and technology are relevant for our ethical or political behavior. Even scant attention to news, popular science magazines, or television makes us realize the difficulty of such an undertaking. Such a task would require an analysis of issues related to stem cell research, therapeutic cloning, genetic enhancement techniques, gene therapy, genetically modified organisms, reproductive technologies, transgenic animals, and many other advances. To try to tackle all of these issues in a single book would prevent an analysis of sufficient depth for any particular issue. Hence, although there are many cases in which biological knowledge has been used to defend claims about moral behavior or policy decisions, I will focus here on three that I believe have special relevance. In chapter 2, I will assess the claim that if genes determine human behavior, then social responsibility is diminished. After a brief introduction to the scientific aspects of cloning in chapter 3, chapters 4 and 5 will explore claims about the morality and immorality of cloning human beings. Finally, in chapters 7 and 8, I will evaluate the defense of moral obligations to seek information about our genetic endowment, to inform family members who might be at risk of carrying harmful genetic mutations, and to avoid bringing seriously ill babies into this world. Such discussion is preceded by an introduction in chapter 6 of some of the current genetic testing techniques and their use.

I consider these three cases because they have received a great deal of attention from academics, the media, and the general public. Furthermore, these cases range from individual moral duties to societal obligations. Additionally, the cases are representative of the kinds of mistakes often present in discussions related to the moral and public policy implications of biomedical science and technology. The evaluation of these sample cases will, hopefully, wake us up from our slumber so that we can begin to pay careful attention to what biology can and cannot tell us about ethical and policy decisions. Not to do so runs the risk of misinforming the public about biomedical science and technology, of promoting inadequate public policies, and of encouraging questionable moral behavior.

Epistemological Problems

Biological information has often been used to support claims about societal responsibility. What is it that we can or cannot know about our social responsibilities by acquiring information about our biological nature? Can we justifiably argue that if particular human traits and behaviors are genetically determined, then social responsibility is diminished? As we will see in chapter 2, some authors have argued that if genes rule, the attempt to change seemingly unfair social systems, educational policies, or institutional arrangements will not succeed, or at least that such transformations would require too great a cost.[2] As examples, I will consider four human traits that have often been presented as biologically determined: intelligence, aggression, addictive behavior, and sexual differences in reproductive strategies. The controversy surrounding these traits and behaviors is about whether they are in fact genetically determined. I will show that this controversy is misconceived. Even if genes "determine" these traits, the conclusion about a diminished social responsibility does not follow.

In chapter 2, I will assess the arguments presented to defend the claim that if particular human traits and behaviors are genetically determined, then social responsibility is diminished. I will show that these arguments contain an epistemological error: the error of assuming that knowledge about particular biological traits and behaviors gives us knowledge of the value of such traits and behaviors. It is erroneously presupposed that judgments about whether human traits and behaviors are good or bad, appropriate or inappropriate, suitable or unsuitable, can be made independently of the environment in which such traits and behaviors are expressed. In the case of human beings, this environment includes social values, arrangements, and institutions. Thus, even if we were able to obtain accurate information about one part of this system, the biological part, such information would be incomplete when making value claims. Hence, claims that societal moral responsibility diminishes with the truth of biological determinism are mistaken because they incorrectly presuppose that our social environment is irrelevant when evaluating human traits and behaviors that matter to us.

Analyses of our social responsibility that make this epistemological mistake might promote an inadequate understanding of human biology. In this case, they offer support to the idea that human biology alone is relevant when evaluating the value of certain human traits. Similarly, such evaluations might encourage social policies that unjustly burden particular individuals or groups. No less important, assessments of social responsibility suffering from this epistemological error might encourage individuals to be unconcerned with a careful evaluation and, when needed, transformation of particular social institutions.

Scientific Problems

That a misunderstanding of scientific knowledge has grounded the defense of moral and political claims is hardly a novel idea. Infamous are the cases of, for example, the eugenics movement produced by a misunderstanding of human genetics and evolution, or the careless use of X-rays prompted by a misconception of radiation.[3] Of course the fact that the present mistake is not new in kind does not make it any less questionable or insidious. Given the importance that Western society in particular concedes to science, it seems extremely important to ensure a correct understanding of scientific issues.

Unfortunately, many of the discussions related to biological knowledge and its consequences for moral and public policy matters are plagued by a misconception of the role of genes in human biology.[4] In many of these discussions, genes are presented as the main determinants of human traits, behaviors, or diseases. These discussions often disregard relationships between genes, epigenetic effects (i.e., modifications that affect an organism's phenotype without changing the DNA sequence), the influence of the cellular environment on gene expression, and the effects of environmental and social factors. Of course everybody is eager to claim that such other factors are important, even essential for understanding how genes work. Nonetheless, these other biological, environmental, and social factors appear as background conditions and thus become immaterial to the discussion. But, which factors are presented as background conditions is obviously not imposed by the world; rather, it is a pragmatic choice made by investigators. Although genes are thus not seen as both necessary and sufficient conditions for traits or diseases, they are believed to be more necessary or nearly sufficient than other biological, environmental, and social factors.

Because of the enormous attention the possibility of cloning human beings has received, it seems important to evaluate whether similar misconceptions about the role of genes in human biology are present in that particular debate. In chapter 4, I will show that such has been the case. Many of the arguments supporting or rejecting the cloning of human beings suffer from a misunderstanding of human biology. These arguments are grounded on the incorrect assumption that our human genome is not just what makes us humans, but what gives us dignity, what makes us who we are as individuals. Hence, many have opposed cloning by arguing that this practice can produce serious psychological harms to the cloned child, such as a lack of a sense of individuality or unique identity. Likewise, some authors have supported cloning human beings by maintaining that this technique would enable some individuals to clone a person who has special meaning to them. For instance, people who have lost a loved one could clone the person who died. Presumably this would help people replace, for example, a dead son or daughter with a "copy." Nevertheless, nothing in current biological knowledge can justify the belief that a clone would be a copy of the dead son or daughter. Significantly enough, and not surprisingly, those who make these arguments fervently reject biological or genetic determinism. But, as mentioned before, if this rejection were taken seriously, then it would not be possible to argue, for example, that cloning would rob the clone of a sense of individuality.

Current debates about the moral obligations created by our ability to obtain genetic information about ourselves and our offspring have received no less attention. In part due to the success of some of the genetic testing programs available, and in part due to the increased knowledge of molecular genetics, a number of authors have argued that under certain circumstances we have a duty to seek information about our genetic endowment, an obligation to inform those members of our family who might also be at risk of carrying harmful genetic mutations, and a duty not to bring seriously ailing babies into this world.

Autonomy and beneficence are often used as grounds to defend these moral obligations. To make autonomous decisions, one ought to gain information about one's genetic endowment. Acquiring information about our genetic endowments enhances our autonomy because such information allows us to shape our lives according to our own will. Similarly, it is argued that our duty to prevent unnecessary harm requires that we inform family members who might be at risk of being affected by genetic disorders, and that we avoid bringing into the world children who might suffer from a serious disability or disease.

However, as I will show in chapter 7, justifying these moral obligations requires more than appeals to autonomy and beneficence. It must also presuppose several things about our technical capabilities and about human biology. First, it must assume that we are able to obtain knowledge about our genetic makeup and that of our offspring. This requires technological advances in the form of, for example, genetic testing. But as we will see, this ability is not sufficient to justify these moral obligations. Second, it must presuppose that genes alone are relevant when predicting our future health status. Suppose, for example, that there is a genetic test that would inform people of whether they are at risk of suffering from a disease X. For people to argue that we have a moral obligation to perform such a test, they must presuppose not just that the test is available, but also that the test accurately predicts our future health state. This means not only that the test has to be reliable in its identification of the genetic material, but that the identification of such genetic material is sufficient to inform us accurately about our health status in the future. Unless one presupposes that there is a high correlation between having a particular gene and having a particular disease, it would seem odd to claim that a person has a moral obligation to obtain such genetic information, to inform others, or to avoid bringing a child with such a gene into the world.

Although at present there are a variety of available genetic tests, and although in many occasions these tests accurately identify a particular gene, for most diseases and disabilities that affect humans, these tests cannot reliably predict our future health or lack thereof. To affirm that they can is to presuppose incorrectly that genes alone can determine our health status. And as far as we know, they cannot do so for several reasons. First, although there are no questions about the fact that genes are factors that contribute to the state of our health, they are not the only factors. Genes, in the majority of the cases, do not act alone but with other genes. Moreover, where some genes might be somehow "defective," other genes might be able to take over their function. Furthermore, genes are expressed in cells that are part of human organisms. How the cellular environment affects the expression of some genes is far from clear, but it is certainly clear that it does. A genetic test would tell us nothing about these other causal influences.

Second, the concepts of health and disease as applied to humans are all but uncontroversial. It is clear, however, that health and disease cannot be assessed by simply looking at genes, not even at genes in the context of whole organisms. Take, for example, the case of allergic reactions to a substance that is present, and in great quantities, only in highly industrialized societies. Even if such allergic reactions were mainly determined by having some genetic material, we would be hard-pressed to call this a disease or disorder. Indeed, we would be hard-pressed to be concerned with it at all were we living in a nonindustrial society. Or consider the case of some Italian speakers who have neurological markers for dyslexia but show no learning impairment as compared with English-speaking dyslexics who have a much more difficult time learning to read because of the complexity of the English language. It seems, then, that to evaluate human diseases, disabilities, or disorders and their effects, one must take into account the ecological and social environment in which human beings grow and develop. Human biology is not independent of where we live and how we live. Any defense of moral obligations, which are presumably grounded on biological knowledge, cannot neglect all of the aspects mentioned here.

This scientific mistake, that is, the inadequate perception about the role of genes in human disease and disorders, might contribute to the development of public policies that promote unnecessary uses of genetic testing and screening, preimplantation diagnoses, or genetic selection of embryos for complex disorders. Questionable views about the role of genes might lead to the limitation of research efforts aimed at other, broader-ranging, and perhaps more successful responses to human diseases, such as primary prevention, institutional transformation, or environmental regulations. Explicitly, they might lead us to look for medical/technological solutions to problems that may have other more appropriate solutions. The idea that genes are the main contributors of human diseases and human traits could also encourage a lack of individual self-care and an attitude of unavoidable fate. For example, some studies indicate that individuals who find they are at risk of heart disease through the use of a genetic test feel the disease is more inevitable than if they find out about the risk by other clinical tests. Moreover, by analyzing human traits, behaviors, or diseases as if they were exclusively the result of the play of our genes, and as completely independent of our social life, we can contribute to the discrimination against already disadvantaged groups of our society. We can also miss the opportunity to improve the aspects of our social, political, and legal systems that need to be improved. It seems obvious, then, that to the extent that we believe genetic determinism is not just false but misleading and dangerous, philosophers, bioethicists, and scientists all need to be careful not to present their discussions in ways that both presuppose and contribute to the ideology of genetic determinism.

An analysis of the epistemological and scientific mistakes present in many of our discussions about biology and its implications for our morality and public policy can bring to our attention the fact that a correct comprehension of the workings of biology indicates that claims about human biology cannot support, for example, conclusions about diminished social responsibility. Such an understanding also shows that human biology cannot underpin claims about the morality or immorality of human cloning and cannot be used to justify obligations to obtain information about our genetic endowment, to inform other family members, or to avoid bringing affected children into the world. By analyzing the soundness of the epistemological and scientific information that grounds these moral claims, we can learn what biology can tell us. If we ground our moral claims on relevant biological knowledge, and if we can show that such knowledge cannot support such claims, then we must either change such moral beliefs or ground them on something other than current biological knowledge. This is something that biology can tell us.

Moral Problems

The cases I analyze in this book involve not only epistemological and scientific assumptions that are problematic but also questionable conceptions of morality. This is, then, a problem with our ethical theories. Thus, although in these cases the science underlying the moral claims may be correct, and although the consequences of the scientific knowledge may be right, still the trouble is that such consequences do not have the moral import ascribed to them. In particular, discussions about human cloning and the moral obligations that presumably follow from our newly found ability to acquire genetic knowledge involve a troublesome presupposition about the irrelevance of the social context in which these moral claims appear.

In order to evaluate this questionable conception of morality, I will assume the correctness of the scientific and epistemological assumptions underlying the arguments defending the morality of cloning human beings or the presumed obligations to obtain information about our genetic makeup, to inform our family members, and to avoid bringing particular children into this world.[5] Of course, as with claims about the absurdity of genetic determinism, affirmations about the need to take into account the social context are plentiful. But, as with the case of genetic determinism, were one to assess seriously the social context in which these moral obligations would take place, one would be forced to recognize that such a defense is inadequate. The social context is here again taken as a background condition for practical purposes. But this is a decision made by scholars rather than one imposed on us by the world, and thus it is open to evaluation.

As we will see in chapter 5, when important social factors are taken into account, even under the assumption of the correctness of the scientific data presented, still we cannot legitimately conclude that it is moral to clone human beings. Not surprisingly, from reading analyses of this technology, one might get the impression of a society where the most serious and pressing problems are the requests of those who want to replace their dead loved ones, where genetic disease is the main cause of preventable death, where infertility is the most urgent problem, where individuality is threatened, where one of the worst things that can happen to children is that their parents have too many expectations because of their genetic makeup, and where resources are all but limited. And in a world where these were the most pressing sources of human suffering, the kind of debate about human cloning that is occurring now would probably make perfect sense. But such is not really the world we live in. Hence, if we examine the social context in which cloning would take place, we see that the arguments that have been offered to defend the morality of human reproductive cloning are unsuccessful.

Similarly, in chapter 8, I will show that the justification of the putative moral duties related to the ability to obtain genetic information fails even if we grant that genetic tests do give highly predictive information about future health status. This is so because, as in the case of reproductive cloning, the defense of such obligations presupposes the legitimacy of defending moral duties in a decontextualized and abstract fashion. Such an understanding of moral obligations, I will show, is questionable for several reasons. First, it is unhelpful to the real human beings to which these moral obligations presumably apply. Second, and more importantly, neglecting the social and political context in which people make moral decisions runs the risk of reinforcing or furthering injustices in an already unjust world. When we consider important contextual issues such as access to genetic testing and education, disproportionate burdens on women because of their reproductive and social roles, and concerns about resource allocation, we can see that the defense of these presumed moral obligations is far from justifiable.

By analyzing the particular conception of morality that grounds moral and policy claims about cloning and genetic knowledge, we will see that an accurate understanding of the science is not, by itself, sufficient to decide moral questions regarding the use of these technologies. This should put to rest concerns that the arguments here presented might be promoting a poorer understanding of morality by emphasizing the need for adequate scientific knowledge. Hopefully the evaluation of the moral conception present in these discussions will call attention to the fact that, although biology can tell us a great deal about moral and public policy issues, it cannot tell us everything we need to know in order to properly evaluate moral claims related to biomedical science and technologies.

SOME CAVEATS

As mentioned earlier, this book does not attempt to give a comprehensive response to the question of what biology can and cannot tell us about morality and public policy issues. Given my claim about the need to consider not just human genetics but other biological, environmental, and social factors, to endeavor to give such a comprehensive answer would be very difficult, if for no other reason than time and space constraints. I hope that this work will serve, however, as a step to make current discussions of biotechnologies richer and more complex than they often are. The decision to include epistemological, scientific, and moral concerns is also part of a desire to not present the consequences of biological knowledge for morality and public policy in simple ways. I understand that in many cases such a task is impractical, in part because of disciplinary issues, and in part because of space constraints imposed by many professional journals. Nonetheless, I believe it is important to call attention to the interrelatedness of questions that arise from the research, implementation, and use of biological science and technology.

Of course this is not to say that only the particular issues I consider here are important. My decision about what to include has to do with what I have perceived as neglected in the debates about biotechnologies. Because there have been many powerful critiques of genetic determinism, I thought that a general discussion of epistemological and scientific problems related to determinism was unnecessary. Concerns about genetic causality,[6] genetic reductionism,[7] the concept of "gene,"[8] and the notion of genes being "for" phenotypes[9] are also important in many debates about biotechnologies, but discussions of these issues are also plenty, and thus I do not directly address them. Similarly, I do not deal here with the naturalistic fallacy, in part because this has also been the topic of numerous books and articles, and in part because I, like many others, think this fallacy is rarely, if ever, committed by anyone.[10] Thus, as should be clear by this introductory chapter, this book is not directed toward answering the question of whether we can reasonably expect biology to tell us anything at all about morality or public policy. My assumption is that biology can and must be taken into account when discussing ethical and political concerns, but biology should not be our exclusive concern in such discussions. This book should be seen as a modest attempt to show that an assessment of what biology can and cannot tell us when evaluating moral claims and policy decisions is extremely timely and imperative. It should be seen as an effort to show that, by carefully analyzing some of the epistemological and scientific mistakes present in these discussions, we can learn a little about what biology can tell us. Further, by putting our moral discussions in their proper context—the social context in which human beings would be bound by such beliefs—we can also learn a little about what biology cannot tell us.

NOTES

1. For critical analysis of the optimistic embracing of the genetic revolution, see R. Hubbard and E. Wald, *Exploding the Gene Myth* (Boston: Beacon Press, 1997); D. Nelkin and S. Lindee, *The DNA Mystique: The Gene as a Cultural Icon* (New York: W. H. Freeman, 1985). For some accounts that express a sanguine approach to the new genetics, see, for example, W. Gilbert, "A Vision of the Grail," in *The Code of Codes: Scientific and Social Issues in the Human Genome Project*, ed. D. J. Kevles and L. Hood, 83–97 (Cambridge, MA: Harvard University Press, 1992); P. Kitcher, *The Lives to Come: The Genetic Revolution and Human Possibilities* (New York: Simon & Schuster, 1997); J. Watson, *A Passion for DNA: Genes, Genome, and Society* (New York: CSHL Press, 2000).

2. See chapter 2 for references.

3. See, for example, D. Paul, *The Politics of Heredity* (Albany: State University of New York Press, 1998); D. Kevles, *In the Name of Eugenics: Genetics and the Uses of Human Heredity* (New York: Knopf, 1995); A. Stewart and G. W. Kneale, "Radiation Dose Effects in Relation to Obstetrics, X Ray and Childhood Cancer," *Lancet* 1, no. 7658 (1970): 1185–87; J. S. Walker, "The Controversy over Radiation Safety. A Historical Overview," *Journal of the American Medical Association* 262, no. 5 (1989): 664–68.

4. See chapters 4 and 7 for references.

5. See chapters 5 and 8 for references.

6. See, for example, F. Gifford, "Understanding Genetic Causation and Its Implications for Ethical Issues in Human Genetics," in *Mutating Concepts, Evolving Disciplines: Genetics, Medicine, and Society*, ed. R. Ankeny and L. Parker, 109–25 (Dordrecht, the Netherlands: Kluwer Academic Publishers, 2002); E. Sober, "The Meaning of Genetic Causation," in *From Chance to Choice: Genetics and Justice*, A. Buchanan,

D. Brock, N. Daniels, and D. Wikler, 347–70 (New York: Cambridge University Press, 2000); S. Oyama, *Evolution's Eye: A Systems View of the Biology-Culture Divide* (Durham, NC: Duke University Press, 2000), especially chap. 5; J. S. Robert, "Interpreting the Homeobox: Metaphors of Gene Activation in Development and Evolution," *Evolution and Development* 3, no. 4 (2001): 287–95; L. Gannett, "What's in a Cause?: The Pragmatic Dimensions of Genetic Explanations," *Biology and Philosophy* 14, no. 3 (1999): 349–74; J. Dupré, *The Disorder of Things: Metaphysical Foundations of the Disunity of Science* (Cambridge, MA: Harvard University Press, 1995), 171–217; C. Cranor, "Genetic Causation," in *Are Genes Us? The Social Consequences of the New Genetics*, ed. C. Cranor, 125–41 (New Brunswick, NJ: Rutgers University Press, 1994).

7. See, for example, S. Sarkar, *Genetics and Reductionism* (Cambridge, UK: Cambridge University Press, 1998); J. Dupré, *The Disorder of Things*, 85–167; C. K. Waters, "Why the Anti-Reductionist Consensus Won't Survive: The Case of Classical Mendelian Genetics," in *PSA 1990*, ed. A. Fine, M. Forbes, and L. Wessels, 125–39 (East Lansing, MI: Philosophy of Science Association, 1990); P. Kitcher, "1953 And All That: A Tale of Two Sciences," *Philosophical Review* 93, no. 3 (1984): 335–76; J. A. Fuerst, "The Role of Reductionism in the Development of Molecular Biology: Peripheral or Central?" *Social Studies of Science* 12, no. 2 (1982): 241–78; R. Dawkins, *The Selfish Gene*, 2nd ed. (New York: Oxford University Press, 1989); M. Ruse, "Reduction in Genetics," in *PSA 1974*, ed. R. S. Coen et al., 633–51 (Dordrecht, the Netherlands: Reidel, 1976); K. F. Schaffner, "Reductionism in Biology: Prospects and Problems," in *PSA 1974*, ed. R. S. Coen et al., 613–32 (Dordrecht, the Netherlands: Reidel, 1976).

8. See, for example, L. Moss, "Deconstructing the Gene and Reconstructing Molecular Developmental Systems," in *Cycles of Contingency: Developmental Systems and Evolution*, ed. S. Oyama, P. Griffiths, and R. Gray, 85–97 (Cambridge, MA: MIT Press, 2001); P. J. Beurton, H.-J. Rheinberger, and R. Falk, eds., *The Concept of the Gene in Development and Evolution* (Cambridge, UK: Cambridge University Press, 2000); P. Griffiths and E. Neumann-Held, "The Many Faces of the Gene," *BioScience* 49, no. 8 (1999): 656–62; C. K. Waters, "Genes Made Molecular," *Philosophy of Science* 61, no. 2 (1994): 163–85.

9. See, for example, L. Moss, *What Genes Can't Do* (Cambridge, MA: The MIT Press, 2002); J. M. Kaplan and M. Pigliucci, "Genes 'for' Phenotypes: A Modern History View," *Biology and Philosophy* 16, no. 2 (2001): 189–213; K. Sterelny and P. Kitcher, "The Return of the Gene," *The Journal of Philosophy* 85, no. 7 (1988): 339–61; R. Dawkins, *The Extended Phenotype: The Long Reach of the Gene* (New York: W. H. Freeman, 1982).

10. See, for example, F. Fukuyama, *Our Posthuman Future: Consequences of the Biotechnology Revolution* (New York: Farrar, Straus and Giroux, 2002), especially chap. 7; A. Rosenberg, *Darwinism in Philosophy, Social Science, and Policy* (Cambridge, UK: Cambridge University Press, 2000), 118–36; P. Woolcock, "The Case Against Evolutionary Ethics Today," in *Biology and the Foundation of Ethics*, ed. J. Maienschein and M. Ruse, 276–306 (Cambridge, UK: Cambridge University Press, 1999); G. Munévar, "The Morality of Rational Ants," in *The Naked Truth: A Darwinian Approach to Philosophy*, 131–47 (Aldershot, UK: Ashgate, 1998); R. J.

Chapter Two

Biological Explanations and Social Responsibility

One need not be an expert to be aware of how claims about human biology have been used to justify the status quo or to promote public policies that discriminate unfairly against people. From claims about women's inability to think rationally because of their biology, to the use of brain size to justify discrimination against African Americans, to the eugenics experiments performed by the Nazis, history is filled with examples of the use of biological claims to support alarming public policies.[1] It is no wonder, then, that heated debates emerge anytime new claims about human biology appear. The technological successes produced by knowledge of genetics have certainly increased people's interests and concerns for the implications that such knowledge has for our view of who we are, what moral duties follow from such knowledge, and how public policies might be changed. Nonetheless, such interests and concerns are far from novel. From Aristotle over two millennia ago, to Galton in the nineteenth century, to our present-day evolutionary psychologists and genetic behaviorists, biological explanations—albeit in different forms—seem to be the obvious answer to our quest for understanding anything that has to do, not just with human diseases or human health, but also with our social arrangements and institutions. These people all appear to have derived lessons about society, politics, and ethics from what they took, and take, to be a biological understanding of human nature. The successes in molecular genetics and the Human Genome Project have only increased our expectations for these kinds of answers. Although we thought our fate was in the stars, we now seem to believe that our destiny is in our genes.

For some, comprehending human biology is essential to solving the problems that loom ahead, from genetic diseases to social unrest.[2] These people

15

believe that if we want public policies to solve problems such as criminal behavior, alcoholism, learning disabilities, schizophrenia, or war, we ought to
pay attention to the evolutionary history of humankind, to its biological nature. Failing to do so is a recipe for disaster. As Steven Pinker says,

> It is not just that claims about human nature are less dangerous than many peo-
> ple think. It's that the *denial* of human nature can be *more* dangerous than peo-
> ple think. This makes it imperative to examine claims about human nature ob-
> jectively, without putting a moral thumb on either side of the scale, and to figure
> out how we can live with the claims should they turn out to be true.[3]

Others, however, remind us of the unjust public policies promoted by social Darwinists or eugenics experiments, and policies sanctioned by Germany
or the United States, all performed under the umbrella of biological knowledge.[4] Critics of biological explanations of human nature insist that many of
these explanations presuppose an unsustainable biological determinism.[5]
They maintain that biological explanations of human behavior are reduction-
ist; that they overestimate the role of genes; and that to look for adaptations
in every one of our emotions, behaviors, and abilities is simply to misunderstand how natural selection works.

In spite of these disagreements, most critics and supporters of biological
explanations seem to concur on one issue: that if biological determinism
were correct, then we would be exempt from critically analyzing and maybe
transforming our social practices and institutions.[6] In the words of Dorothy
Nelkin,

> Evolutionary principles imply genetic destiny. They de-emphasize the influence
> of social circumstances, for there are natural limits constraining individuals. The
> moral? No possible social system, educational or nurturing plan can change the
> status quo. Evolution, defined as an eternal principle "writ large," becomes a
> way to justify existing social categories and to deflect critical examination of the
> powers underlying social policy.[7]

So, trying to change social institutions or social policy would fail, or, at
least, trying to do so would require too great a cost. As Edward Wilson explains, "Human nature is stubborn and cannot be forced without a cost."[8] Biology then turns out to be a way to justify existing social institutions and to
relieve societal guilt. If genes were the cause of crime, addiction, racism, depression, and so on, public policies and social institutions would be irrelevant
to those problems. If so, we would need not evaluate these policies and institutions. Neither would we need to feel responsible for such problems, as our
collaborative efforts are not at all implicated in their existence.

Critics accuse those who use biology to explain every possible human trait of presupposing the truth of biological or genetic determinism. As Hilary and Steven Rose say,

> Within this framework new forms of biological determinism have arisen . . . modern and more sophisticated forms, powered by the vast expansion of biological, and particularly genetic, knowledge and technology. This biological determinism takes two apparent antithetical forms. On the one hand, it claims our biology is our destiny, written in our genes by the shaping forces of human evolution through natural selection and random mutation. This biological fatalism is opposed by Promethean claims that biotechnology, in the form of genetic engineering, can manipulate our genes in such a way as to rescue us from the worst of our fates.[9]

Thus these critics try to debunk genetic determinism, and with it presumably the claims made by sociobiologists, evolutionary psychologists, and behavioral geneticists.[10] All the while, proponents of these disciplines agree that genetic determinism is false and has dangerous implications for our society. Evolutionary psychologists and genetic behaviorists, however, deny that the biological explanations of human nature that they put forward are based on the assumption that human behavior is genetically determined. According to Pinker, "Nor any other sane biologist would ever dream of proposing that human behavior is deterministic, as if people commit acts of promiscuity, aggression, or selfishness at every opportunity."[11]

The debate on the social consequences of biological or genetic determinism suffers from a misunderstanding of the workings of human biology and genetics. Claims about the devastating consequences of biological or genetic determinism for our social responsibility presuppose that our biological traits and behaviors can be evaluated outside of the environmental, social, and political contexts in which such traits and behaviors are expressed. However, genetic predispositions have to be expressed as phenotypic traits—that is, as observable physical or behavioral characteristics that result from the interplay of genes and environments—before we can evaluate whether these characteristics are good or bad things. But many human phenotypic attributes diverge in value according to the social environmental context in which they are expressed. For instance, homosexuality, assuming for the sake of the argument that this is a genetically determined trait, can be very problematic in societies that place great value on the connection between sexual acts and reproductive ones, but it would be unlikely to raise much concern in social environments where such a connection is irrelevant. Controversial biological explanations about human intelligence, sex differences, aggression, or addictive behavior deal with characteristics that must be evaluated in social contexts. Such social contexts are not fixed. They have changed over human history, and there seem to be no reasons to believe that we cannot change them again in order to pursue worthy moral goals such as, for example, equality or fairness. Hence, it appears that to argue that we must give up social responsibility because certain important traits result from our biological endowment is not only a philosophical mistake but also an exercise in bad biology.

The focus of this chapter is to show that an epistemological mistake is responsible for the common belief that the correctness of biological or genetic determinism implies the end of critical evaluation and reform of our social institutions. That is, these arguments commit the error of assuming that knowledge about particular biological traits and behaviors is sufficient to give us knowledge of the value of such traits and conducts. They erroneously presuppose that human traits and behaviors are intrinsically good or bad, appropriate or inappropriate, suitable or unsuitable. I argue here that eliminating this epistemological mistake, and thus understanding the role of biology properly, would make us realize that what constitutes a problem for those who are concerned with social responsibility is not the fact that particular behaviors might be genetically determined, but the fact that our value system and social institutions create the conditions that make such behaviors questionable. And they do so by making a causal contribution to a trait or behavior's development and by informing judgments concerning the trait or behavior's desirability. Thus, a proper understanding of human biology requires that we pay attention to the social context in which human traits and behaviors are expressed.

SOME CAVEATS

It is hardly surprising that most people who are concerned with the relationships between genetics and human traits and behaviors agree that genetic determinism is false. The kind of determinism they tend to reject is what Kaplan calls the "complete information" and the "intervention is useless" versions of genetic determinism.[12] The first version affirms that our genes dictate everything about us. The second strand asserts that for traits that have a genetic component, intervention is useless. There is, however, another version of genetic determinism that is presupposed by many who do not see themselves as genetic determinists. In this version, traits with even partial genetic etiologies are best understood as primarily genetic, and only through directed intervention can we avoid or control the expression of genes for such traits.[13] Even when genes are not determining, then, they are perceived as more necessary or more fundamental than other biological and environmental counterparts.[14]

My discussion includes both the strongest and the moderate versions of genetic determinism.

BIOLOGY AS DESTINY

Because of an epistemological misconception about the role of biology in human life, many people seem to believe that if we can explain particular human traits and behaviors in biological terms, then we have good reasons to ground claims about our lack of social responsibility. Hence, if we believe that our genes determine intelligence, then it would seem that there is no justification for continuing to invest societal resources in improving the intelligence level of those who cannot do better. Similarly, if aggression is a genetic predisposition, then social circumstances seem irrelevant to the production of criminal behavior. If alcoholism or other kinds of addiction—to drugs, to food, or to sex—are in our genes, assessing current social practices that might be responsible for promoting these behaviors would appear futile. And if the differences between men and women are an inextricable part of human nature, then claims of gender equality would seem entirely misguided.[15]

These common beliefs, however, are groundless. They erroneously presuppose that we can evaluate human traits and behaviors by referring only to their biological origins or determination. But, as I said earlier, biological traits have to be evaluated in the environment where they are expressed. Because humans are social creatures, their environment includes not only the natural environment, but also social values, arrangements, and institutions. Biological or genetic determinism can thus, at most, give us partial information about the value of particular human traits and behaviors. If a strong genetic determinism is presupposed, then the belief that social responsibility must be relinquished ignores the social contingency of judgments about the desirability of particular traits or behaviors. If, on the other hand, a weaker version of genetic determinism is assumed, then the idea that social responsibility diminishes disregards the causal contribution of the human social environment to the development of such a trait or behavior, as well as the eventuality of judgments concerning its appropriateness. Hence, claims that societal moral responsibility decreases with the truth of biological determinism are mistaken because they incorrectly presuppose that the social environment is irrelevant, either as a causal contributor to a biological trait or behavior or as a contributor to the judgments about such a trait.

In what follows, I will analyze some of the human traits that have been presented implicitly or explicitly as being genetically determined. Note that my point is not to deny that such traits may be biologically determined.[16] I want to argue that the consequence—that is, that it prevents social responsibility— does not follow. Nor am I claiming that those who offer biologically based explanations for these traits believe that they are biologically determined; on the contrary, as I have said earlier, most of them go to painstaking lengths to argue that they are not claiming such a thing.[17] And they do so because, along with their critics, they consider biological determinism not only scientifically but also morally dubious. The human traits that I will evaluate are the following: intelligence, aggression, addictive behavior, and reproductive strategies in men and women.

Intelligence

In spite of the ease with which we discuss it, intelligence, like the proverbial happiness, is difficult to define. Most of us believe that to be intelligent is a good thing, even if each of us understands what it means to be intelligent in different ways. Moreover, the difficulties surrounding the definition of intelligence are compounded by difficulties related to its measurement and quantification. Whatever it might be, however, we are clear that the more you have of it, the better off you will be. The value of intelligence in our society seems to be unquestionable.

It is not surprising, then, that one of the most contentious claims in the dispute over biological explanations has been the assertion that intelligence is inheritable and to a high degree immutable. These claims are normally associated with evidence that allegedly shows that Caucasians are more intelligent than African Americans.[18] Of course, because of the long history of racial conflict and discrimination, these assertions have been especially controversial in the United States. Presumably the differences in intelligence explain inequalities in society: those who are more intelligent tend to be more successful. The high rates of poverty that afflict Western societies in general, and the United States in particular, are determined more by intelligence than by socioeconomic background. Thus, the argument often goes, using social resources to enrich the education of those at the low end of the cognitive distribution is inefficient.[19] It doesn't matter how many resources are provided for these children; their cognitive abilities and their chances of succeeding in our world are not going to improve significantly. Society is then doubly exculpated. First, we don't need to feel responsible for those individuals who do not "make it." After all, no matter how much was done for them, it would not improve their chances. And second, we do not need to be critical of social practices and institutions that appear to increase the chances that some people are not going to be able to "make it." After all, if you are smart enough, you will, sooner or later, overcome these problems.

But does this consequence of abdicating social responsibility follow from the claim that intelligence is genetically determined? I will argue here that it does not. The argument about the genetic determination of intelligence relies on several assumptions. For instance, it presupposes that heritabilities can be compared between different populations. It also presupposes that the notion of human intelligence is uniform, that intelligence can be described by a single number, and that it is capable of ranking people in some linear order. Similarly, it assumes that intelligence is clearly measured by IQ tests, that it is genetically based, that it is immutable, and that there is a black-white differential in IQ. Further, it also presupposes that those who do well on IQ tests tend to occupy high positions in academia, in the business world, and in politics, and that those who do poorly on IQ tests tend to occupy lower positions on the social ladder. This means that there will be a correlation between IQ and wealth and social status: the higher your IQ, the better your chances are to become wealthy and respected; the lower your IQ, the better your chances are to be poor. Now, if IQ is immutable, then special education programs, which try to increase people's intelligence levels, will not do the job. At this point, it seems that all is lost for those concerned with social equality.

Let us ignore, for the sake of the argument, some of the serious problems that have been pointed out about these presuppositions.[20] Imagine, then, that these assumptions are correct. Nevertheless, the claims about what follows socially from the fact that intelligence is genetically determined are mistaken. They presuppose that it is possible to evaluate human intelligence outside of the social environment in which humans live, independently of an evaluation of how such a social environment informs people's judgments about what is good or bad. Thus, such claims can be correct only if we believe that not only intelligence but also our social context, value system, and political institutions are immutable. But such a presupposition is quite questionable. If these social factors and value judgments are changeable, then there are a few things that a society committed to social justice could do. For example, we could decide to pay better those who now occupy low-wage jobs (presumably because they are not smart enough to do any better); this might go a long way toward improving the lives of people who are less smart. Or we could change the kinds of things we value most. Instead of valuing wealth, we could begin to value things such as the ability to care for others, to keep promises, to enjoy other people's company, to be concerned with our environment, or to be able to grow a garden. Or we could begin to realize that academics, CEOs, and scientists do not succeed only because of their personal efforts and because of their intelligence. They would not be able to do much without the help of the people who pick up their garbage, who clean their houses, or who harvest the vegetables and fruits they eat. Changing our value system and social institutions could certainly go a long way toward improving the lives of those who happen to be less intelligent. Those changes could go as far as to wrest importance from the fact that some people might be more intelligent than others. For example, imagine that tomorrow a new scientific study shows conclusively that the strength required to be able to become a first-rate wrestler is hereditary. No matter how much one trains, no matter how many social programs we implement to help people become better wrestlers, the result would be the same: those who have the necessary genetic material could succeed, and those who don't should give up.[21] Now, it is unlikely that in the present social context this would cause too much of a stir. Why? Because in our value system, wrestling does not rank high: if you can and want to do it, it might be a good thing; if you can't and don't, there are many other things, maybe even better ones, that humans can do with their lives. The value we give to wrestling is dependent on our social values, not only on whether it is biologically determined. Similarly, the value we give to human intelligence depends on particular social values. Thus, in a social context in which intelligence, as it is normally understood in these debates, is not seen as a necessary condition to having a meaningful life, those who happen to be less intelligent would still have chances. There may be many other traits they do have that might be valued by our society. If my argument is correct, then genetic determinists' claims about intelligence make critical evaluation of our social institutions and values even more important and necessary when one is committed to social justice.

Aggression

In 2003, U.S. residents of age twelve or older experienced over 24 million crimes. Nearly 18.6 million were property crimes, 5.4 million were crimes of violence, and 185,000 were personal thefts.[22] In this context, claims about the naturalness of aggression are bound to be troublesome. Of course statements asserting that human beings are innately aggressive and violent are not new.[23] Nor are theories trying to show that criminal behavior has a biological or genetic basis.[24] Well-known is the case of the alleged strong positive correlation between criminal behavior and individuals with XYY karyotypes.[25]

As with intelligence, many fear that claims about the innateness of aggression can be used to divert attention from oppressive economic and social conditions that might be recognized as contributors to violence.[26] Thus, explaining aggression and violence as genetic predispositions can justify our blaming particular individuals while overlooking their social circumstances. According to these critics, in a society like ours, extremely concerned with efficiency and happy to embrace anything that keeps us from critically evaluating our social practices and institutions, the idea that aggression is innate and immutable could translate into an abandonment of rehabilitation plans, efforts to help disadvantaged groups, and welfare programs.[27] Although it is the case that our society often has been quick to embrace biological explanations of human traits and behaviors as a way to justify questionable public policies, we must ask whether such justification is legitimate. If, as I mentioned before, the evaluation of biological traits and behaviors cannot be done in a vacuum, then knowing that human beings are innately aggressive says nothing about the value of such a trait. Therefore, it is hard to see how it can follow from the claim that aggression is genetically determined in humans that we are exempt from carefully and critically analyzing our social values and institutions to see how they are contributing to our value judgments of aggression.

Let us assume, then, that aggression is genetically determined. This could mean several things. For example, it could mean that aggressiveness is a defining trait of human nature.[28] But if this is the case, it seems that the exculpation of a critical analysis of our social institutions does not follow. Aggression might be judged as a good thing in certain social contexts. We don't seem to have a problem with aggressive businesspeople or with using aggression as a drive to improve ourselves. Thus, aggression in itself is not the problem; what is the problem is the direction that aggression might take, for instance, toward antisocial behavior. If this is so, we might be able to analyze critically our social practices and institutions to see how they are directing aggression. We might decide to promote changes in our social structures in ways that direct aggression toward good or noncontentious ends.

To say that aggression is genetically determined could also mean, not that all human beings are genetically determined to be aggressive, but that some, because of their genetic makeup, are.[29] Thus, particular individuals have genes that increase their chances of committing criminal or antisocial acts. It would seem, then, that under the dual assumption that these traits are immutable and that whoever has these particular genes will be aggressive, the claim that we are exempt from evaluating and transforming our social institutions is justified. But, as before, unless we presuppose that our value system and social context are unchangeable, and thus we neglect the social environment's causal contribution to the development of this behavioral trait, as well as the eventuality of judgments about its desirability, we are not exonerated from a critical analysis of our social organization. Criminal or antisocial conduct is only one way among others to express aggressiveness. If this is the case, then aggressiveness can be redirected, and therefore we need to know what kinds of social schemes might be able to promote aggression that would be judged desirable. The differences in antisocial behavior between the United States and

Western European countries, for example, might tell us something about how different social practices encourage or discourage certain kinds of aggressive behavior. Maybe in Europe most people with these "aggressive genes" are encouraged to play soccer.

If, rather than presupposing a strong version of genetic determinism, a weaker one is assumed, then the responsibility of a community to be critical of its institutions and practices would be even more pressing. This is so because, in a weaker strand of genetic determinism, genes might be sufficient conditions for being aggressive, but they might not be necessary conditions. For instance, people might have genes "for" curly hair, but fashion trends can obviously do a lot to encourage people without such genes to become curly haired. Thus, even if someone does not have the "aggressive genes," a person might become aggressive if faced with certain social contexts, such as poverty, discrimination, abuse, and so on. If so, then we ought to be concerned with the kinds of social structures that might promote aggression even in those not genetically determined to be so.

Addiction

Addiction to different substances such as alcohol, nicotine, or illegal drugs has also often been explained in biological terms. Some scientists have claimed that there are particular genes responsible for addiction.[30] They maintain that the sequencing of the human genome will help us to understand the biology of addiction by allowing us to identify genes that contribute to individual risk and those through which drugs cause this behavioral trait. Presumably, individuals who have these genes are prone to become alcoholics, heavy smokers, or drug addicts.[31]

Critics have seen these kinds of biological explanations as problematic. But, as before, the problem does not lie with the fact that the trait or behavior might be genetically determined. To assert that it does presupposes that we can make judgments about biological traits independently of the social and political context in which they are expressed. Obviously, in a world where dangerous addictive substances are not easily attainable, the fact that people might have genes that make them more likely to become addicts to such substances would probably raise little concern. In this context, this behavioral trait would be unlikely to be a subject of study. If this alleged biological trait spurs debate, it is because in our society, drugs, alcohol, and nicotine are easily available. Also, claims about the genetic origin of addiction appear problematic in a society where social arrangements and institutions are such that they permit, if not outright encourage, unjust discrimination against individuals, lack of employment and educational opportunities, and unavailability of medical care. It is clear that these social conditions, even if not the cause of addictive behaviors, cannot help those who might presumably be at risk of becoming addicts.

Moreover, to proclaim that if addiction were genetically determined this would affect social responsibility ignores the social environment's causal contribution to this trait or behavior. Addictive substances are necessary for someone with a particular genetic endowment to become an addict. So, even if some people have genes that determine that they will become alcoholics, the addiction will happen only if the individuals in question consume alcohol. The alcohol and tobacco industries' use of biological explanations to claim that their products are not the problem, but that the individual's genes are, are thus misguided.[32] In fact, it is not obvious why the accountability of the industry selling these products would be different depending on the genetic status of addiction. Provided that people are adequately informed about the consequences of using these products, whether they become addicted solely because they use them or because they use these products and have a biological predisposition to addiction seems irrelevant. The product is a necessary condition for the addiction, even if it is not sufficient. The industry's responsibility to provide information and to make this information available does not diminish. On the contrary, it seems that, if anything, it would increase. Suppose, for example, that we discover that people who are naturally blond have a higher risk of becoming alcoholics than do people with other hair colors. And suppose that the alcohol industry, knowing this, presents an aggressive advertisement campaign inciting blond people to drink. Presumably we would not take as exculpatory a defense that says, "had the person not been naturally blond, the alcohol would have been just fine." Or, most likely, we would not exculpate the tobacco industry for possible deaths if they directed a campaign toward people with pulmonary problems.

What is the case for the industry is also the case for society at large. Although we might want to fault the individual, and often do so, and avoid a painstaking evaluation of our social structures, genetic determinism cannot legitimize such an attitude. Some social arrangements are more conducive to encouraging people to use drugs and alcohol in abusive ways. Assuming that our society considers addictive behavior a problem for the well-being of the individual in particular and the good of the community in general, we have a responsibility to analyze our social institutions and practices to see if, and in what ways, they might be promoting that behavior. Consider for example the high rate of alcoholism among Native Americans who reside on reservations. There is some evidence that for genetic reasons Native Americans do not metabolize alcohol very well.[33] But the fact that liquor stores are placed at the border of reservations is a social phenomenon that contributes to the rate of alcoholism in these populations. Add to this, other social phenomena such as the high level of unemployment, the limited control over natural resources on their lands, and the historically restricted access to education, and we find that social institutions and values play a large role in the rates of alcoholism for Native Americans.

As we can see, even if it is the case that some people, because of their genetic makeup, are prone to becoming addicts, it still might be the case that particular social contexts make people less likely to choose drinking, smoking, or drugs. We might want to promote these kinds of social contexts. Similarly, if we find that certain social arrangements, such as those that restrict employment or educational opportunities or that discriminate against people for unfair reasons, tend to make people more likely to make a choice to drink, smoke, or consume drugs, we might decide to change such arrangements. If we are concerned with people's and society's well-being, we can also encourage educational courses that would help people to be aware of their possible risks, as well as medical facilities and programs that would help those who become addicts. As in prior cases, genetic or biological determinism by itself does not justify an uncritical acceptance of our value system and our social institutions.

Sex Differences

Biological explanations have also been prominent in matters of sex differences.[34] As the recent debate over Harvard president Lawrence Summers's suggestion that innate differences between men and women help explain the lack of top-level female professionals in science and engineering shows, such explanations are also highly controversial.[35] Of course, trying to explain the differences between women and men in biological terms is far from a new scientific approach. However, current sociobiologists and evolutionary psychologists have been especially attentive to this issue.[36] According to these disciplines, the different physical constitutions of the sexes provide different strategies for maximizing their fitness through the reproduction of the largest possible number of offspring. The psychological characteristics that make males effective reproducers are likely to be different from, and to some extent in conflict with, those that make females effective. If this is so, then genes that are successful in transmitting copies of themselves into future generations are likely to be ones that have different physiological and psychological effects in the male and female bodies that they help to construct. One would expect, then, for males and females to have different natures and behaviors, and therefore social roles.

Differences in reproductive strategies between men and women are related to differences in parental investment. Men produce millions of sperm a day and, without much investment, can theoretically father offspring with different women daily. Women, however, have a much greater investment in each offspring because their eggs are much larger than the sperm, they normally produce only one egg per month and about four hundred in their lifetimes, and they usually produce no more than one offspring a year. Furthermore, women are the ones who gestate the fetus in their bodies, and the ones who will nurse and most likely will take care of the baby for years. Thus, women contribute a larger proportion of their reproductive potential and employ more time and energy than men. As a result of these differences, women and men have developed different reproductive strategies: women are selective and choosy, and it is in their best interests to hold back until they can identify males with the best genes; on the other hand, it pays for men to be aggressive, hasty, promiscuous, and undiscriminating.

Sociobiologists and evolutionary psychologists maintain that modern women have developed, from their successful ancestors, wisdom and prudence about the men with which they consent to mate. Because women seek resources in a partner, and because resources might not be directly discerned, women have developed traits that allow them to identify qualities in men that signal likely possession or future acquisition of resources, such as intelligence, ambition, and older age. But having resources is not enough. Women need to make sure that the mate they choose would be willing to invest those resources in them and their offspring. Thus, women seek traits that indicate this willingness to invest, such as dependability and stability, love cues, and positive interactions with children. But, of course, characteristics such as good health, size, strength, bravery, or athletic ability are also important to modern women because tall, strong, athletic men offered ancestral women protection, and we are descendants of women who chose this way.

Men, on the other hand, supposedly have greater desire for short-term mating than do women. The reproductive advantages to ancestral men as a consequence of short-term mating would have been an increase in the number of offspring produced. Modern men, thus, express greater desire for a variety of sex partners, are more undiscriminating, have more sexual fantasies, and have a larger number of extramarital affairs. However, because many ancestral women required reliable signs of male commitment before consenting to sex, men who failed to commit would have failed to attract many women. Thus, it was beneficial for ancestral men to seek long-term partnerships. Also, this would have increased certainty in paternity and would have enhanced the survival and reproductive success of his children. But, to be reproductively successful, ancestral men had to marry women who could bear children. Because women's reproductive capacities cannot be observed directly, men have evolved standards of attractiveness that represent clues to women's ability to reproduce, such as youth and health as indicated by clear skin, full lips, symmetrical features, absence of sores and lesions, white teeth, and a small ratio of waist to hips. Also, because of the problem of paternity uncertainty, men have evolved preferences for fidelity in women. Thus, modern men want physically attractive, young, sexually loyal wives who will remain faithful throughout their lives.

Of course, most sociobiologists and evolutionary psychologists are careful to point out that the fact that sexual differences are biological does not mean that they are unchangeable.[37] They might mention, though, that to try to create gender equality will only come at serious costs. As Edward O. Wilson argues, "There is a cost, which no one can yet measure, awaiting the society that moves either from juridical equality of opportunity between the sexes to a statistical equality of their performance in the professions, or back toward deliberate sexual discrimination."[38] Human beings' behaviors might be malleable, but not that much.

Critics of the kind of biological explanations just discussed worry that the alleged natural coyness of women, and the intense male competition and aggression it presumably produces, serve to justify the prominence of men in positions of power; social structures that encourage and allow for unequal opportunity and access to resources for women; as well as male violence against women, rape, and adultery. Thus, they worry that if sociobiologists and evolutionary psychologists are correct about the genetic origin of sex differences, then no amount of political initiative, social spending, or political disruption will change the essentially unequal relationships between men and women. If sexual differences are genetically determined, then the fight for sexual equality seems misguided. And, again, a critical evaluation of our social structures appears to be little more than futile.

As in the cases of intelligence, aggression, and addiction, it seems far from clear that an abdication of our critical and transforming abilities follows from the fact that sexual differences might be genetically determined and unchangeable. Again, as before, it seems that such an abdication might follow only if we presuppose that our current social context is immutable. But, as I said earlier, such an assumption is quite questionable. Let us then presuppose, for the sake of the argument, that sexual differences can be explained mainly in biological or genetic terms. Let us, then, accept that women will always tend to prefer high-status, older men who would provide for their offspring, and that men will tend to prefer younger, beautiful women who will be faithful wives and nurturing mothers. Do these purportedly biologically determined desires doom women to a life of subordination, and men to a life of domination? I think not. Let us begin with, for example, women's alleged nurturing desires and tendencies. Imagine that we believe that encouraging women's caring abilities is good for women in particular and for society in general. This would be so because women would do what they desire to do, and our youngsters would be nurtured and cared for. We can conclude, and many have done so, that if these nurturing tendencies cannot be changed, then it is better for women to remain at home taking care of their children. But, of course, there are other options. For example, we can also propose that all working facilities have appropriate day-care services. Thus, mothers would be able to work outside of the house at the same time that they have a chance to satisfy their alleged biological desires. We can also encourage social changes directed toward offering women part-time jobs that allow these women, without overworking them, to advance in their careers and take care of their families. Or we can promote systems of tenure and promotion that do not penalize women for doing what it is presumably in their nature to do, to have children and care for them. Similarly, we can support policies that offer benefits, such as health care for them and their children, tax benefits, and so on, to women who decide to stay home and take care of their families. Furthermore, although it might be the case that women are biologically equipped to take care of their offspring, this should not be an impediment to promoting social changes that would encourage men to be more nurturing. As we can see, critical evaluation of our societal structures seems far from unnecessary, even when we presuppose that women are naturally inclined to nurture and care for their babies and that they are the ones better equipped to do so. As before, this biological trait says nothing at all unless we present it in the social context in which we live. It is precisely this social context that makes this particular trait controversial because of the historical discrimination against women.

The same can be said about the different mating strategies employed by men and women. That men desire young, beautiful wives who will be faithful and caring, and that women desire older men with resources is a problem in societies where these kinds of alleged biologically evolved traits systematically disadvantage women and favor men. In a society where men tend to have the resources and women tend to be dependent on men for their own support and that of their children, these biological desires appear highly problematic. An evolutionary desire for younger women is a problem in a society where older women lack communal and personal resources and are continuously disregarded and devalued. Women's alleged desire for older, richer men is a serious problem in a society where that might be the only way for many to acquire status and opportunities. The same can be said about men's desire for beautiful women. The problem is not that they want beauty but that the lack of opportunities faced by women might compel the female population to see beauty as the only way to have access to social and individual resources.

Alas, these presumed biological desires cannot require that we abdicate critical evaluation and reform of our social institutions. Thus, even if these mating strategies have been naturally selected and therefore are not easy to overcome, still there is much we can do in the direction of gender equality. For example, we could promote laws, regulations, or social institutions that give women equality of opportunity in access to high-paid, high-status jobs. If women had the economic resources and the social assistance necessary to support themselves and their children, it would seem irrelevant for the social status of women that they still might desire older, richer men, even if as a matter of fact they do so because of their biology. Moreover, if women have financial resources and communal support for themselves and their children, the fact that men have a tendency to be unfaithful would be less significant for women's equality. Unfaithfulness in a partner certainly could be a problem for a particular woman, but not different from having friends who are dishonest or siblings who might deceive her.

SOME OBJECTIONS AND REPLIES

Someone might object that my arguments here fail because I have ignored the possibility that human values might also be genetically determined. If our value systems were biologically determined, critics might say, then we would have serious difficulties changing them. Thus, to propose a change in values or a transformation of our social institutions to better achieve social justice would be not just naive, but futile.

Although this objection raises an important issue, it is questionable. This is so because, even if we accept that humans' value systems are genetically determined, the content of expression need not be. It might certainly be the case that we have evolved a natural moral sense out of the requirements of our species and that there are particular emotional responses that lead to the formation of moral ideals in more or less uniform ways.[39] But, of course, this does not require humans to be born with innate abstract moral principles or ideas. Consider again my discussion of aggression. Even if valuing aggression is genetically determined, how aggression is expressed might be a matter of social choices. We can encourage people to channel aggression into sports rather than antisocial behavior. Similarly, even if we agree that competition is a biologically determined tendency and that we are also biologically determined to value this tendency, what we compete for, or the manner of competition, might depend on societal decision. A society where people compete to improve each other's well-being would likely be quite different than another where people compete for wealth. A society where the competition is directed toward improving our moral character would probably be unlike one where individuals compete to obtain wealth.

Critics might also argue that my evaluation has neglected the possibility that our social institutions are the result of our biologically determined values. In this view, particular systems are doomed to fail because they go against human biology. For instance, utopias that are grounded on the malleability of humans would not succeed.[40] If this were the case, then any attempt to promote institutions that go against our human nature or to change those that follow from such nature would be a failure.

The idea that our social arrangements and institutions might be determined by our biology seems to fly in the face of history. If one presumes that human nature, including our values, is genetically determined and that our social arrangements and institutions are the direct result of such nature, then one must explain the variety of social systems and arrangements that have formed part of human life throughout our history. It is difficult to see, for example, how a feudalist, a democratic, and a totalitarian system could all be the result of the same biologically determined values. Thus, such claims seem to contradict the historical evidence. Of course this is not to say that our biological nature presents no constraints to the kinds of social arrangements into which humans might enter. Quite likely, some social institutions—for example, those that encourage freedom—will more likely contribute to individuals' well-being than will social arrangements that force people into slavery. Nonetheless, given the historical evidence, it seems difficult to argue that such biological constraints determine a particular social system, rather than allow for a variety of them.

CONCLUSION

For much of human history, biology has been used to justify the status quo. It is not surprising, then, that renewed interest in biological explanations appears worrisome for those who are concerned with social justice. However, to criticize biological or genetic determinism by pointing out that if it is correct then individuals are not responsible for critically evaluating and maybe transforming our institutions is unhelpful. Such criticisms commit an epistemological mistake. They misunderstand the role of biology in human life. They disregard the fact that these biological traits are contentious because of the social context in which they appear, rather than because of some intrinsic property they might have. Also, they can disregard the causal contribution of humans' social environment to the development of particular traits or behaviors. Thus, a critical evaluation and transformation of our values and social arrangements appears more, not less, required were it the case that particular human traits and behaviors are genetically determined.

Finally, it is important to notice that I am not denying that critics of biological explanations have well-grounded fears because social institutions are, and have been, using biology to defer responsibility for social injustices. People have used, and continue to use, biological explanations as a way to oppress some and unjustly benefit others. Neither am I denying that our society tends to see biology as destiny. Nor am I claiming that genetic determinism is true. My point is simply to call attention to the fact that an abandonment of social responsibility does not follow from the truth of biological or genetic determinism. To maintain that it does is to erroneously presuppose that our social context has nothing to do with what makes these biological explanations wearisome. That is, it misunderstands the workings of biology. Biological traits are rarely intrinsically good or bad; judgments about their desirability depend on the environmental context in which they are expressed, which in the case of humans includes social and political contexts. Critics, in their attempt to block the use of biology as a way to justify what are morally problematic social policies, have not been sufficiently attentive to this point. And this is seriously questionable in a world where more and more alleged evidence is presented of the biological origins of many of our traits and behaviors. By affirming that such biological origins legitimize the abandonment of social responsibility, not only might we be giving more power to biology than it really has, but we might also be missing the opportunity to improve the aspects of our social, political, and legal systems that need to be improved, and we might be opening the doors to those who want to use biology to ground unjust public policies. If I am right, a correct understanding of human biology can inform arguments supporting the need for critical evaluation of our value system and of our social and political institutions. Social transformation can go a long way toward curtailing the possible limitations that biology might impose on human beings.

NOTES

1. See, for example, D. Kevles, *In the Name of Eugenics: Genetics and the Uses of Human Heredity* (New York: Knopf, 1995); N. Tuana, *The Less Noble Sex: Scientific, Religious, and Philosophical Conceptions of Woman's Nature* (Bloomington: Indiana University Press, 1993).

2. See, for example, J. Cartwright, *Evolution and Human Behavior* (Cambridge, MA: The MIT Press, 2000), 341–43; R. Thornhill and C. T. Palmer, *A Natural History of Rape: Biological Basis of Sexual Coercion* (Cambridge, MA: The MIT Press, 2000), 199; M. Daly and M. Wilson, "Evolutionary Psychology and Marital Con-

flict," in *Sex, Power, Conflict: Evolutionary and Feminist Perspectives*, ed. D. M. Buss and N. M. Malamuth, 9–28 (New York: Oxford University Press, 1996); D. Buss, "Sexual Conflict: Evolutionary Insights into Feminism and the 'Battle of the Sexes,'" in *Sex, Power*, 296–318; R. Wright, *The Moral Animal: Why We Are the Way We Are; The New Science of Evolutionary Psychology* (New York: Pantheon, 1994), 10–14.

3. S. Pinker, *The Blank Slate* (New York: Penguin, 2003), 139. Italics in the original.

4. D. Paul, *The Politics of Heredity: Essays on Eugenics, Biomedicine, and the Nature-Nurture Debate* (Albany, NY: State University of New York Press, 1998); R. Hubbard and E. Wald, *Exploding the Gene Myth* (Boston: Beacon Press, 1997), 13–22; D. Kevles, *In the Name of Eugenics*.

5. See, for example, S. Rose, "Moving on from Old Dichotomies: Beyond Nature-Nurture towards a Lifeline Perspective," *British Journal of Psychiatry*, suppl. 40 (2001): S3–S7; T. Benton, "Social Causes and Natural Relations," in *Alas, Poor Darwin: Arguments against Evolutionary Psychology*, ed. H. Rose and S. Rose, 249–72 (New York: Harmony Books, 2000); R. Lewontin, *The Triple Helix* (Cambridge, MA: Harvard University Press, 2000); S. Oyama, *Evolution's Eye: A Systems View of the Biology-Culture Divide* (Durham, NC: Duke University Press, 2000); T. Shakespeare and M. Erickson, "Different Strokes: Beyond Biological Determinism and Social Constructivism," in *Poor Darwin*, 229–48; S. Rose, "Escaping Evolutionary Psychology," in *Poor Darwin*, 299–320; R. Bleier, *Science and Gender: A Critique of Biology and Its Theories on Women* (New York: Pergamon Press, 1985); P. Kitcher, *Vaulting Ambition: Sociobiology and the Quest for Human Nature* (Cambridge, MA: The MIT Press, 1985); S. J. Gould, *The Mismeasure of Man* (New York: W. W. Norton, 1981).

6. See, for example, R. Thornhill and C. T. Palmer, *History of Rape*, 107–11; A. Buchanan, D. Brock, N. Daniels, and D. Wikler, *From Chance to Choice: Genetics and Justice* (Cambridge: Cambridge University Press, 2000), 24–26; D. Buss, *Evolutionary Psychology: The New Science of the Mind* (Boston: Allyn & Bacon, 1999), 18–19; M. Rothstein, "Behavioral Genetic Determinism: Its Effects on Culture and Law," in *Behavioral Genetics: The Clash of Culture and Biology*, ed. R. Carson and M. Rothstein, 89–115 (Baltimore, MD: Johns Hopkins University Press, 1999); D. Nelkin, "Behavioral Genetics and Dismantling the Welfare State," in *Behavioral Genetics*, 156–71; D. Nelkin and S. Lindee, *The DNA Mystique: The Gene As a Cultural Icon* (New York: W. H. Freeman, 1985); R. Wright, *Moral Animal*, 345–63; P. Kitcher, *Vaulting Ambition*.

7. D. Nelkin, "Less Selfish Than Sacred? Genes and the Religious Impulse in Evolutionary Psychology," in *Poor Darwin*, 27.

8. E. O. Wilson, *On Human Nature* (Cambridge, MA: Harvard University Press, 1978), 147.

9. H. Rose and S. Rose, "Introduction," in *Poor Darwin*, 5.

10. For works criticizing these disciplines, see, for example, J. Dupré, *Human Nature and the Limits of Science* (New York: Oxford University Press, 2001); J. Kaplan, *The Limits and Lies of Human Genetic Research* (New York: Routledge, 2000); M. Midgley, *Beast and Man: The Roots of Human Nature* (London: Routledge, 1996); H. Rose and S. Rose, eds., *Poor Darwin*; A. Fausto-Sterling, *Myths of*

Gender (New York: Basic Books, 1985); P. Kitcher, *Vaulting Ambition*; S. J. Gould, *Mismeasure of Man*.

11. S. Pinker, *Blank Slate*, 112–13.

12. See J. Kaplan, *Limits and Lies*, 11–12.

13. J. Kaplan, *Limits and Lies*, 12.

14. L. Gannett, "Tractable Genes, Entrenched Social Structures," *Biology and Philosophy* 12, no. 3 (1997): 403–19.

15. I am going to presuppose throughout this chapter that what theorists of human behavior, both from biological and environmentalist approaches, mean by "behavior" is clear and uncontroversial. Of course, I do not believe that to be the case. For discussions on the complexity of defining what behaviors are, see, for example, H. Longino, "Behavior as Affliction: Common Frameworks of Behavior Genetics and Its Rivals," in *Mutating Concepts, Evolving Disciplines: Genetics, Medicine, and Society*, ed. R. Ankeny and L. Parker, 165–87 (Dordrecht, the Netherlands: Kluwer Academic Publishers, 2002); J. Dupré, *Human Nature*; S. Rose, "The Poverty of Reductionism," in *Thinking about Evolution: Historical, Philosophical, and Political Perspectives*, vol. 2, ed. R. Singh et al., 415–28 (Cambridge, UK: Cambridge University Press, 2001); E. Balaban, "Behavior Genetics: Galen's Prophecy or Malpighi's Legacy?" in *Thinking about Evolution*, 429–66; R. Amundson, "Against Normal Function," *Studies in the History and Philosophy of Biological and Biomedical Sciences* 31, no. 1 (2000): 33–53.

16. See references in note 10 for arguments denying that these and similar traits are genetically determined.

17. For a discussion of what the denial of genetic determinism might mean, see Kaplan, *Limits and Lies*, chap. 2.

18. See, for example, A. Jensen, *The g Factor: The Science of Mental Ability* (Westport, CT: Praeger Publishers, 1998); R. Herrnstein and C. Murray, *The Bell Curve: Intelligence and Class Structure in American Life* (New York: Free Press, 1994).

19. See previous note for references.

20. J. Kaplan, *Limits and Lies*, chaps. 3 and 4; B. Devlin, S. Fienberg, D. Resnick, and K. Roeder, eds., *Intelligence, Genes, and Success: Scientists Respond to The Bell Curve* (New York: Springer-Verlag, 1997); S. Fraser, ed., *The Bell Curve Wars: Race, Intelligence, and the Future of America* (New York: Basic Books, 1995); R. Jacoby and N. Glauberman, eds., *The Bell Curve Debate* (New York: Times Books, 1995); S. Rose, R. Lewontin, and L. Kamin, *Not in Our Genes: Biology, Ideology, and Human Nature* (New York: Pantheon, 1984); S. J. Gould, *Mismeasure of Man*.

21. Of course this does not mean that someone can be a successful wrestler without any training, or that someone who trains but does not have the "appropriate" genes will not be better at it than someone who does not have the "wrestling genes" and does not train at all.

22. U.S. Department of Justice, Office of Justice Programs, Bureau of Justice Statistics, *Criminal Victimization 2003*, NCJ 205455 (Washington, DC: Department of Justice, 2004), http://www.ojp.usdoj.gov/bjs/pub/pdf/cv03.pdf (accessed 31 Jan. 2005).

23. See, for example, D. Barash, *The Whispering Within* (New York: Harper & Row, 1979); L. Tiger and R. Fox, *The Imperial Animal* (New York: Holt, Rinehart, and Winston, 1971); L. Tiger, *Men in Groups* (New York: Vintage Books, 1970).

24. See A. Raine, *The Psychopathology of Crime: Criminal Behavior as a Clinical Disorder* (New York: Academic Press, 1997).

25. During the mid-sixties, a study done by geneticist Patricia Jacobs and colleagues reported that XYY males were present with a higher-than-expected frequency in institutions for the criminally insane. This affirmation led to the unsubstantiated belief that XYY males were predisposed to criminal behavior. See, for example, J. Kaplan, *Limits and Lies*, 96–98; R. Hubbard and E. Wald, *Gene Myth*, 104–7.

26. L. B. Andrews, "Predicting and Punishing Antisocial Acts: How the Criminal Justice System Might Use Behavioral Genetics," in *Behavioral Genetics*, 116–55; D. Nelkin and L. Tancredi, *Dangerous Diagnostics: The Social Power of Biological Information* (Chicago: University of Chicago Press, 1995).

27. L. B. Andrews, "Antisocial Acts."

28. S. Pinker, *Blank Slate*, 306–36; D. Buss, *Evolutionary Psychology*, 278–311; R. Wrangham and D. Peterson, *Demonic Males: Apes and the Origins of Violence* (Boston: Houghton Mifflin, 1996); L. Tiger and R. Fox, *Imperial Animal*; L. Tiger, *Men in Groups*.

29. M. J. Gotz, E. C. Johnstone, and S. G. Ratcliffe, "Criminality and Antisocial Behaviour in Unselected Men with Sex Chromosome Abnormalities," *Psychological Medicine* 29, no. 4 (1999): 953–62; L. F. Jarvik, V. Klodin, and S. S. Matsuyama, "Human Aggression and the Extra Y Chromosome: Fact or Fantasy?" *American Psychologist* 28, no. 8 (1973): 674–82.

30. D. E. Comings et al., "The Additive Effect of Neurotransmitter Genes in Pathological Gambling," *Clinical Genetics* 60, no. 2 (2001): 107–16; E. Duaux, M. O. Krebs, H. Loo, and M. F. Poirier, "Genetic Vulnerability to Drug Abuse," *European Psychiatry* 15, no. 2 (2000): 9–14.

31. T. Burnham and J. Phelan, *Mean Genes* (Cambridge, MA: Perseus Publishing, 2000), 58–82.

32. D. Nelkin, "Behavioral Genetics and Dismantling the Welfare State," in *Behavioral Genetics*, 160.

33. See, for example, T. L. Wall, L. G. Carr, and C. L. Ehlers, "Protective Association of Genetic Variation in Alcohol Dehydrogenase with Alcohol Dependence in Native American Mission Indians," *The American Journal of Psychiatry* 160, no. 1 (2003): 41–46; T. L. Wall et al., "Alcohol Dehydrogenase Polymorphisms in Native Americans: Identification of the ADH2*3 Allele," *Alcohol and Alcoholism* 32, no. 2 (1997): 129–32.

34. A. Fausto-Sterling, *Sexing the Body: Gender Politics and the Construction of Sexuality* (New York: Basic Books, 2000); N. Tuana, *Less Noble Sex*.

35. S. Dillon, "Harvard Chief Defends His Talk on Women," *New York Times*, January 18, 2005.

36. S. Pinker, *Blank Slate*, 337–71; C. Badcock, *Evolutionary Psychology: A Critical Introduction* (Malden, MA: Polity Press, 2000), 149–87; J. Cartwright, *Evolution*, 92–156, 212–60; L. Mealey, *Sex Differences: Developmental and Evolutionary*

Strategies (San Diego: Academic Press, 2000); R. Thornhill and C. T. Palmer, *History of Rape*, 31–52; D. Buss, *Evolutionary Psychology*, 97–185; R. Wright, *Moral Animal*, 33–151; M. Daly and M. Wilson, *Sex, Evolution, and Behavior* (Boston: Willard Grant, 1983); E. O. Wilson, *Human Nature*, 121–48.

37. R. Thornhill and C. T. Palmer, *History of Rape*; D. Buss, "Sexual Conflict," 306; E. O. Wilson, *Human Nature*.

38. E. O. Wilson, *Human Nature*, 147.

39. See, for example, F. Fukuyama, *Our Posthuman Future: Consequences of the Biotechnology Revolution* (New York: Farrar, Straus and Giroux, 2002), pt. 2; J. R. Richards, *Human Nature after Darwin: A Philosophical Introduction* (London: Routledge, 2000); A. MacIntyre, *Dependent Rational Animals: Why Human Beings Need the Virtues* (Chicago: Open Court, 1999); W. Rottschaefer, *The Biology and Psychology of Moral Agency* (Cambridge, UK: Cambridge University Press, 1998); M. Ruse, *Taking Darwin Seriously: A Naturalistic Approach to Philosophy* (Amherst, NY: Prometheus Books, 1998); M. Nussbaum, "Aristotle on Human Nature and the Foundations of Ethics," in *World, Mind, and Ethics: Essays on the Ethical Philosophy of Bernard Williams*, ed. J. E. J. Altham and R. Harrison, 86–131 (Cambridge, UK: Cambridge University Press, 1995); M. Nussbaum, "Non-Relative Virtues: An Aristotelian Approach," in *The Quality of Life*, ed. M. Nussbaum and A. Sen, 1–6 (New York: Oxford University Press, 1993); Aristotle, *Nicomachean Ethics*, trans. Terence Irwin (Indianapolis: Hackett, 1985).

40. See P. Singer, *A Darwinian Left: Politics, Evolution, and Cooperation* (New Haven, CT: Yale University Press, 2000).

Chapter Three

An Introduction to the Science of Cloning

As we saw in chapter 2, even if particular human traits and behaviors were genetically determined, this would not have the implications for social responsibility that many have feared (or welcomed). Nonetheless, a full understanding of human biology, within its social context, is extremely important. Thus, this chapter aims to get clear about the technology of cloning in ways that will be useful for evaluating the moral issues ensuing from the possible practice of reproductive cloning.

The birth of the sheep "Dolly" in 1996 sparked a public debate about the issue of human cloning. In spite of her much-celebrated birth, Dolly was not the first animal clone to exist. Occasionally, nature itself creates clones when a single fertilized egg cell produces twins or multiple siblings. Similarly, many living organisms, such as some plants, reproduce asexually; that is, they produce other individuals by, for example, budding. When a DNA macromolecule replicates, it creates two identical copies of the original DNA macromolecule. Moreover, bacteria and protozoans multiply simply by dividing in two, as do all animal cells, thus creating identical genetic copies.

Artificial cloning did not appear with Dolly, either. Scientists have been practicing DNA cloning, gene cloning, or molecular cloning since the 1970s, through recombinant DNA technology.[1] This technology, now common in molecular biology labs, allows researchers to transfer a particular DNA fragment from one organism to a self-replicating genetic element such as a bacterial plasmid. In this way, scientists can produce enough identical DNA material for further study. Furthermore, other animals had been artificially cloned much earlier than Dolly was conceived.[2] In 1885, for example, while studying cell differentiation, Hans Driesch was able to produce clones of sea urchins by splitting a two-celled embryo. Both of the cells turned into complete embryos. In the early 1900s, Hans Spemann was able to split two-celled salamander embryos, a more complex organism than a sea urchin, to create two whole new embryos. During the early 1950s, Robert Briggs and Thomas King, also concerned with issues of cell differentiation, devised a new technique, nuclear somatic transfer, to clone frogs. They enucleated—that is, they removed the nuclei—from unfertilized frog eggs and introduced in them the nuclei from early frog embryo cells. They thus were able to obtain normal tadpole clones of the embryos that donated the nuclei. In the late 1960s, John Gurdon produced clones by transferring the nuclei of tadpole intestinal cells into enucleated frog eggs. This produced embryos, tadpoles, and frogs that were clones of the tadpoles that donated the stomach cell nuclei. The first successful mammals cloned, sheep, were brought about by Steen Willadsen in the 1980s. He placed enucleated sheep eggs and cells from sheep embryos side by side. He then subjected the cells to electric shock to fuse them. Some of these newly created clone embryos were then placed into the wombs of surrogate mothers, who gave birth to full-term lambs. Cloning using embryonic cells has since been successful with cows, rhesus monkeys, pigs, goats, and mice.

When Dolly was born, she created such uproar not only because she was a clone—after all, there had been others—but because she was the first mammal cloned from a somatic cell in an adult animal (i.e., a cell other than a germ cell) rather than from embryonic material. This was remarkable because the dogma in biology until that moment was that adult differentiated cells could not be reversed to an embryonic state of totipotency. Totipotent cells are the only ones that can give rise to descendants that may differentiate to form any of the kinds of tissue necessary for an animal to develop properly. The only totipotent cells were thought to be the fertilized egg and the first four or so cells produced by its cleavage or division. After Dolly, cloning of mammals with adult donor nuclei has been successful also with cattle, goats, pigs, cats, mice, rabbits, and horses. Cloning of rhesus monkeys using adult donor cells, however, has not yet been successful.

THE NUTS AND BOLTS OF CLONING

Cloning can be used to refer to several biological processes: cell division, budding, DNA replication, embryo splitting, or nuclear transfer. What all of these have in common is that they create copies of a particular entity. But the ethical, social, or legal implications of these various biological processes are quite different. Our concern in this and the following chapters will be exclusively with what is normally referred to as "reproductive cloning." In this chapter, the discussion will focus particularly on mammalian reproductive cloning. I will thus deal with the intentional creation of born animals that have the same nuclear DNA as another currently or previously existing animal. Each newly created individual is a clone of the original donor of the nuclear genetic material. Notice that in reproductive cloning it is not sufficient that embryos are created. At least some of those embryos must be transferred into a female's womb.

Currently there are two methods to produce live-born mammalian clones: embryo splitting and Somatic Cell Nuclear Transfer.[3] Embryo splitting involves the fertilization of an egg outside a female's body. The union of egg and sperm forms a zygote that divides into two, and then four, identical cells. At this point the cells can be separated and then allowed to develop into separate individuals with identical DNA. Although the animals produced by this method will be genetically identical to each other, they are not genetically identical to their parents. These embryos will have genetic material from two different individuals, the one who contributed the egg, and the one who contributed the sperm. Because individuals born through this method would be no different than naturally born identical twins, which are not morally contentious, I will not deal with this method any further.

Somatic cell nuclear transfer, on the other hand, does not create individuals that have nuclear genetic material derived from two different individuals, as with embryo splitting or normal sexual reproduction. In somatic cell nuclear transfer, the nuclear genetic material derives from only one individual: the donor of the nucleus. The process of reproductive cloning by somatic cell nuclear transfer involves the use of both somatic and reproductive cells, in particular the egg.[4] Animals that reproduce sexually have two different types of cells: germ cells and somatic cells. Germ cells are the reproductive cells that give rise to the gametes: egg or sperm. Mature gametes are haploid; that is, they have only one set of chromosomes. On the other hand, somatic cells are any cells in the body other than germ cells. They are diploid; that is, they have two complete sets of chromosomes. Somatic cells divide by the process of mitosis. Each mitotic division produces two daughter cells from an original parent cell. Germ cells divide to form gametes by a different process called meiosis. In this case, each original cell produces four haploid daughter cells. Egg and sperm come together to produce a new individual. This new individual will inherit genetic material from both parents and thus will be genetically different from each of them.

In somatic cell nuclear transfer, an egg is enucleated under a microscope using a pipette to suction its nucleus.[5] Once this is done, the egg cell has none of its nuclear chromosomes. The chromosomes are then replaced with a nucleus extracted from cultured cells that may have come originally from somatic cells of an embryo, a fetus, or an adult organism. This nucleus, because it is derived from somatic cells, contains a diploid set of chromosomes. Once the new nucleus—sometimes more than just the nucleus—is introduced into the egg cell, it is placed between a couple of electrodes that apply an electric current. The electric shock allows the egg to fuse with the nucleus or donor cell and behave as if it were now a normally fertilized egg cell. The result is a "reconstructed embryo." The reconstructed embryo is normally placed in the oviduct of a temporary recipient. A few days later, the reconstructed embryo, now more developed, is transferred into the oviduct of a permanent recipient. Both temporary and permanent female recipients have been treated with appropriate hormones so that they physiologically respond as if they were pregnant. If a pregnancy occurs, the result will be an organism that has the nuclear genetic material not of two parents, but only of one: the donor of the nuclear chromosomes. All of the embryos produced from the same donor animal will be "twins," and all of them will have the same nuclear genetic material as the donor parent. They will be clones.

Unfortunately, not all reconstructed embryos develop into complete and healthy individuals.[6] On the contrary, the rate of success for mammalian clones is quite low, especially when using adult donor cells, and in most cases their health leaves much to be desired. For example, Dolly was the only success of 277 tries, and she had to be put down by lethal injection in February 2003 because she was suffering from lung cancer and arthritis. She had reached only half the life span of a typical sheep. It is unclear, yet, whether her diseases were the result of the cloning procedure. Only about 1 to 3 percent of reconstructed embryos result in live births. The vast majority of cloned embryos die during embryonic development, and they show a high incidence of developmental aberrations. A variety of abnormalities and defects, both before and after birth, have been observed in reproductively cloned animals. Many of the embryos that are transferred into recipients result in abnormally large fetuses. Others result in dead fetuses. Some of the cloned animals that survive often suffer serious abnormalities, such as deformed limbs, as well us dysfunctional lungs, livers, kidneys, hearts, or immune systems. Clones have also shown placental deficiencies, brain defects, and cardiovascular problems. A variety of organs and tissues are thus affected in cloned animals, and many of these defects cannot be diagnosed or prevented with current technology. Cloning can of course also affect the pregnant female, although documentation and research of this aspect of cloning has been quite sparse. Maternal health problems and mortality have resulted from late-term fetal loss, large fetus size, abnormalities in placental function and development, and excessive fluid accumulation in the uterus.

Several hypotheses have been presented to explain the low success rate of mammalian cloning, but current evidence is still insufficient to corroborate any of them.[7] Some researchers have suggested that the introduction of mitochondria from the donor cell into the egg might be responsible for some of the developmental anomalies found in clones.[8] Mitochondria, one of the several components of animal cells, usually are only maternally derived. During fertilization by sexual reproduction, a very small amount of paternally derived mitochondria enters the egg, but it is eliminated from the zygote, the one-celled embryo formed by the fusion of sperm and egg, during the first few rounds of cellular division. Nuclear transfer procedures, however, can result in the introduction of mitochondria from the donor. As a result, the cloned embryo would have mitochondria derived both from the enucleated egg and from the donor cell.

Another possible reason for the low success rate of mammalian cloning is related to telomere length.[9] Telomeres are sections at the end of chromosomes that prevent them from unraveling. They work as a cap that protects the chromosomes from degradation or fusion with other chromosomes. When a chromosome divides, some of the telomere end is lost. Thus, cells in older individuals have shorter telomeres than do cells in younger individuals. When the shortening gets to a particular point, cellular division ceases. Because somatic cell nuclear transfer procedures employ adult cells or cells that have already passed through several rounds of cellular division, these cells contain chromosomes with shortened telomeres. This has raised questions about the possibility of cloned animals suffering premature senescence, an accelerated aging.

A third hypothesis explaining the numerous clone embryos that do not develop properly has to do with epigenetic processes.[10] Epigenetic changes are modifications of DNA and core histone proteins, that is, proteins that act as the backbone for DNA molecules and that regulate the activity of genes without altering the DNA sequence. These modifications have the ability to turn genes on and off so that genes, in collaboration with proteins and other cellular components, can coordinate early developmental progression and make products necessary for growth of the early embryo. Some researchers have suggested that the removal of some of these epigenetic patterns from the early embryo—for example, by erasing some chemical groups, such as methyl groups, from DNA molecules—is an important step in the reprogramming of DNA for the correct development of the new individual. Erasure of these epigenetic modifications does not occur, or occurs abnormally, in embryos produced by somatic cell nuclear transfer. Other researchers have called attention to the epigenetic process of imprinted genes. Particular chemical groups mark these imprinted genes in the egg or the sperm so that genes behave differently depending on whether they are inherited from the mother or the father. Because in cloned animals the nuclear genetic material derives from only one individual rather than from two different ones, imprinting errors occur. These errors can result in placental and fetal abnormalities and in death.

CLONING OF HUMAN BEINGS

On several occasions, there have been announcements of the creation of cloned human embryos.[11] For example, in 1998, some scientists from South Korea claimed that they had successfully replaced the nucleus of an egg from an infertile woman with the nucleus from a somatic cell. According to this report, researchers had cultivated the clone embryo until it had divided twice, before destroying it. Nobody could corroborate this claim. In 2002, Chinese researchers announced that they had been successful in cloning dozens of human embryos. The result, however, did not appear in any peer-reviewed scientific journal. In 2003, Severino Antinori, a controversial Italian doctor, proclaimed that three women he had treated were carrying clone fetuses in the advanced stages of pregnancy. Nobody has been able to find evidence for these claims. And, early in 2004, South Korean scientists reported in the journal *Science* that they had cloned thirty human embryos, had grown them for a week in the laboratory, and had extracted stem cells for further research.[12] The experiment was not, however, intended to produce pregnancies, and the embryos were not transferred to any women.[13] Thus, in spite of the public commotion that some science fiction books and movies have produced, no evidence exists to date that human reproductive cloning has been successful, or for that matter that it has ever been tried. Moreover, as I said before, reproductive cloning by nuclear transfer with adult cells has not been successful with nonhuman primates.

Before we venture in the next two chapters into the human cloning debate, let me emphasize some points that will become important in the evaluation of this procedure. As was clear with the birth of Dolly, cloning an individual from an adult donor cell does not result in the birth of an already adult organism. Cloned animals have to develop in the uterus of recipient females and eventually become newborns. They do not appear fully grown or with all the experiences of an adult. Like any other human, they will acquire those experiences during their development. Moreover, because human clones must be embryos and fetuses before they can become adults, women must be involved. In any attempt to clone humans, women will be necessary both in order to obtain eggs and in order to carry the pregnancies to term. Similarly, we should not ignore the fact that, at this point, and given the fact that basic mechanisms of development are highly conserved throughout evolution, it cannot be ruled out that the developmental and health problems occurring with other mammalian species could occur in attempts to produce a human clone.

NOTES

1. S. Wright, "Recombinant DNA Technology and Its Social Transformation 1972–1982," *Osiris* 2 (1986): 303–60.
2. See, for example, M. A. Di Berardino and R. G. McKinnell, "The Pathway to Animal Cloning and Beyond—Robert Briggs (1911–1983) and Thomas J. King (1921–2000)," *Journal of Experimental Zoology: Part A, Comparative Experimental Biology* 301, no. 4 (2004): 275–79; J. B. Gurdon and J. A. Byrne, "The First Half-Century of Nuclear Transplantation," *Proceedings from the National Academy of Science of USA* 100, no. 14 (2003): 8048–52; K. Illmensee, "Cloning in Reproductive Medicine," *Journal of Assisted Reproduction and Genetics* 18, no. 8 (2001): 451–67; I. Wilmut, K. Campbell, and C. Tudge, *The Second Creation* (Cambridge, MA: Harvard University Press, 2000), pt. 2.
3. See, for example, Committee on Science, Engineering, and Public Policy and Global Affairs Division, *Scientific and Medical Aspects of Human Reproductive Cloning* (Washington, DC: National Academy Press, 2002), chap. 2.
4. See, for example, Committee on Science, *Human Reproductive Cloning*; I. Wilmut, K. Campbell, and C. Tudge, *Second Creation*, pts. 2 and 3.
5. See note 4 for references.
6. See, for example, D. Humpherys et al., "Abnormal Gene Expression in Cloned Mice Derived from Embryonic Stem Cell and Cumulus Cell Nuclei," *Proceedings of the National Academy of Sciences* 99, no. 20 (2002): 12889–94; J. P. Renard et al., "Nuclear Transfer Technologies: Between Successes and Doubts," *Theriogenology* 57, no. 1 (2002): 203–22; K. Illmensee, "Cloning in Reproductive Medicine."
7. See, for example, I. Wilmut and L. Paterson, "Somatic Cell Nuclear Transfer," *Oncology Research* 13, nos. 6–10 (2003): 303–7; Committee on Science, *Human Reproductive Cloning*, chap. 3; R. Mollard, M. Denham, and A. Trounson, "Technical Advances and Pitfalls on the Way to Human Cloning," *Differentiation* 70, no. 1 (2002): 1–9.
8. See, for example, J. C. St. John et al., "The Consequences of Nuclear Transfer for Mammalian Foetal Development and Offspring Survival: A Mitochondrial DNA Perspective," *Reproduction* 127, no. 6 (2004): 631–41; J. C. St. John, R. Lloyd, and S. El Shourbagy, "The Potential Risks of Abnormal Transmission of MtDNA through Assisted Reproductive Technologies," *Reproductive Biomedicine Online* 8, no. 1 (2004): 34–44.
9. See, for example, P. G. Shiels and A. G. Jardine, "Dolly, No Longer the Exception: Telomeres and Implications for Transplantation," *Cloning Stem Cells* 5, no. 2 (2003): 157–60; D. Betts et al., "Reprogramming of Telomerase Activity and Rebuilding of Telomere Length in Cloned Cattle," *Proceedings of the National Academy of Sciences of USA* 98, no. 3 (2001): 1077–82.

10. See R. K. Ng and J. B. Gurdon, "Epigenetic Memory of Active Gene Transcription Is Inherited through Somatic Cell Nuclear Transfer," *Proceedings of the National Academy of Sciences* 102, no. 6 (2005): 1957–62; H. Tamada and N. Kikyo, "Nuclear Reprogramming in Mammalian Somatic Cell Nuclear Cloning," *Cytogenetic and Genome Research* 105, nos. 2–4 (2004): 285–91; F. Santos and W. Dean, "Epigenetic Reprogramming during Early Development in Mammals," *Reproduction* 127, no. 6 (2004): 643–45; S. Khorasanizadeh, "The Nucleosome: From Genomic Organization to Genomic Regulation," *Cell* 116, no. 2 (2004): 259–72; R. Jaenisch and A. Bird, "Epigenetic Regulation of Gene Expression: How the Genome Integrates Intrinsic and Environmental Signals," *Nature Genetics* 33, suppl. (2003): 245–54.

11. G. Saint-Paul, "Economic Aspects of Human Cloning and Reprogenetics," *Economic Policy* 18, no. 1 (2003): 73–122.

12. W. S. Hwang et al., "Evidence of a Pluripotent Human Embryonic Stem Cell Line Derived from a Cloned Blastocyst," *Science* 303, no. 5664 (2004): 1669–74.

13. G. Vogel, "Human Cloning: Scientists Take Step toward Therapeutic Cloning," *Science* 303, no. 5660 (2004): 937–39.

Chapter Four

Cloning—Or Not—Human Beings

As with many other biomedical technologies, the progress of research on the cloning of mammals has been received with both dismay and excitement because of its implications for human cloning. For some, the cloning of a human being is repugnant, while others see it as a way to solve many of the problems that beleaguer us. Not surprisingly, the cloning debate is plagued with the same problems that we find in the discussion of other biomedical technologies: proponents and opponents of human cloning often present arguments that display their misconceptions about human biology. The purpose of this chapter is to expose those misconceptions.

Many of the arguments rejecting and defending human cloning are grounded on the incorrect assumption that it is our human genome that gives us our individuality and dignity. These arguments attribute a role to the human genome that is not supported by biological facts about how genes operate. This is especially remarkable because, at the same time, authors presenting arguments grounded on this dubious assumption are often careful to point out the falsity of genetic or biological determinism.[1] Needless to say, not all arguments that have been presented in favor of and against human cloning fail because of scientific misunderstandings. Support or rejection of reproductive cloning has been offered in ways that do not presuppose incorrect claims about the role of genes. For example, some have defended cloning as a way to help infertile people who cannot produce their own gametes.[2] This argument will be discussed in chapter 5. Cloning has been rejected because it is seen as a way to explore fundamental mysteries or secrets of life that belong only to God; that is, it is a way to inappropriately "play God."[3] Sometimes, cloning has been rejected because of possible social harms, for example, threats to the stability of the family.[4] I do not attempt to discuss these and similar arguments here. My concern in this chapter is with arguments that are grounded on an incorrect under-standing of biology. As I mentioned earlier, attention to these issues is impor-tant because it can help us to avoid mistakes when defending particular moral or public policy issues in relation to reproductive cloning. Similarly, attention to arguments that incorrectly portray the role of genes in human life is neces-sary because it can assist us in eliminating inaccurate and sometimes danger-ous views about genetic determinism.

ARGUMENTS AGAINST CLONING

Article 1 of UNESCO's 1997 "Universal Declaration on the Human Genome and Human Rights" states that "the human genome underlies the fundamental unity of all members of the human family, as well as the recognition of their inherent dignity and diversity." Article 11 of the same declaration insists, "Practices which are contrary to human dignity, such as reproductive cloning of human beings, shall not be permitted."[5] In 1998, the World Health Assem-bly affirmed "that cloning for the replication of human individuals is ethically unacceptable and contrary to human dignity and integrity."[6] In the same year, the Council of Europe declared, "Any intervention seeking to create a human being genetically identical to another human being, whether living or dead, is prohibited." The council clarifies that "the term human being 'genetically identical' to another human being means a human being sharing with another the same nuclear gene set."[7] In 2002, the President's Council on Bioethics de-clared "that cloning-to-produce-children ought to be legally prohibited. . . . Such cloning, even if it could be done without the risk of defects or deformi-ties, treats the child-to-be as a product of manufacture, and is therefore incon-sistent with a due respect for the dignity of human beings."[8] In spite of the agreement by national and international bodies about cloning being contrary to human dignity, what human dignity really is or how cloning hinders it is far from clear.[9] These claims, however, all seem to presuppose that an individual's unique genome is somehow related to our having, or not having, dignity.

This association of human dignity, or lack thereof, with reproductive cloning cannot be related to the fact that cloning is an artificial means of re-production. International and national bodies did not hurry to declare other ar-tificial means of reproduction, such as in vitro fertilization (IVF) and related technologies, as contrary to human dignity. On the contrary, many national committees embraced these new assisted reproductive technologies as an ap-propriate way to help infertile people.[10]

So, what is it about reproductive cloning that makes this practice inconsis-tent with human dignity? Maybe it is the fact that reproductive cloning eliminates the need for the union of sperm and egg. In all other reproductive technologies, sperm and eggs are still the players in the formation of an embryo. However, as we saw in the previous chapter, nuclear somatic transfer does not call for sperm. Animal experiments have shown us that although the enucleated female egg is still necessary, the nucleus of any other cell will allow us to produce a new being. Certainly, this suggestion that the male gamete provides something more than just the biological material necessary for the formation of a new being is far from new. The idea is already present in Aristotle's works, for example, almost 2,500 years ago. For Aristotle, the male sperm was the cause of change and development, while the female gamete appeared as merely the supplier of the material needed for growth. Of course there is no scientific basis to believe that the role of the sperm in the production of a new being is any more or less than the role of the egg. So, the idea that Dolly was somehow a lesser sheep, or that children born through cloning would lack human dignity because they can be born without the participation of a male gamete, seems preposterous. Imagine, for instance, that a cloned embryo is implanted in a woman's womb and grows into a child. Suppose also that the only person who knows that the embryo was cloned dies before anyone else is informed. We would be now facing a case where the child, because it is a clone, presumably lacks human dignity, but no one would know it. If the creation of the clone were without health dangers, then this child, biologically, would grow into the same sort of creature that we are. This scenario suggests that human dignity in not a biological characteristic passed on through sperm.[11]

Although the lack of a male gamete is something that sets cloning apart from other reproductive technologies, it is not the only thing. Another important difference between cloning and other assisted reproduction techniques is the fact that the cloned entity presumably would have the same genome as the donor of the genetic material. By contrast, in IVF and related technologies, the offspring is genetically different from its parents. But what exactly does it mean to say that the cloned entity would have the same genome as the donor? The genome is normally defined as the totality of genetic information carried by a cell or organism. But we know that the genetic information of a cell is not only contained in its nucleus. Mitochondria also contain genetic information.[12] Mitochondria are membrane-enclosed organelles employed for the production of energy. They are essential for the survival of cells and hence of multicellular organisms. These organelles also have DNA material. In mammals, for example, the mitochondrial DNA contains about 16,500 base pairs. As we saw in chapter 3, mitochondrial DNA in mammals and other organisms is maternally inherited; that is, such DNA comes from the female gamete. It is true that in mammalian cells the mitochondrial DNA makes up less than 1 percent of the total cellular DNA. However, an embryo in which these genomes carry a deleterious mutation cannot survive. And we know now of several inherited diseases that are caused by mutations in mitochondrial DNA genomes.[13] This is important because, as we saw in chapter 3, in reproductive cloning the enucleated egg provides the cytoplasmatic material of which mitochondria are a component. Thus, although the nuclear DNA of the clone would be the same as the nuclear DNA of the donor, the total genome would not. So, unless the enucleated egg and the nucleus to be fused come from the same female donor, it is incorrect to say that a clone has the same genome as the donor of the nuclear DNA.

Of course, someone might object that the amount of mitochondrial DNA is so small that it seems reasonable to discount the differences. But this objection is problematic. It seems unclear why the fact that the mitochondrial DNA makes up a very small portion of a cell's DNA would make any such differences any less important. As mentioned before, a mammal with a deleterious mutation in the mitochondrial DNA would not survive. Moreover, suppose that the egg used in the procedure contains mitochondria with a mutation that causes the inherited disease myoclonic epilepsy and ragged red fiber[14] and that the nucleus comes from a different individual, the donor. The human clone would consequently suffer from a disease that does not afflict the donor.

Let us concede, for the sake of the argument, that declarations about the relation between the uniqueness of the human genome and human dignity do not really refer to the totality of the human genome, but only to the nuclear genome. That is, let us discount the contribution of the mitochondrial DNA to the human genome. The claim, then, would be that human cloning is contrary to human dignity because, presumably, it creates two humans with the same nuclear genome. But again, it is unclear what exactly it means to say that a clone has the same nuclear genome as the donor of the nuclear DNA. In the process of cell division necessary for the creation of a new organism, the nuclear DNA must be replicated.[15] One parent cell gives place to two daughter cells, and these in turn give place to more daughter cells. During replication of the nuclear DNA, however, changes in nucleotide sequences can occur. It is true that cellular proofreading mechanisms and error correction machinery make the amount of possible changes quite small, about 1 in 10^9. But, small as this might be, it would be inexact to say that the nuclear genome of the clone and the donor are identical or the same. After all, the difference between suffering, for example, sickle cell anemia or not is due to a change in a single amino acid.[16] Therefore, biologically speaking, it would be more accurate to say that the genomes of a clone and the donor of the nuclear DNA are quite similar, but they are certainly not identical.

If we cannot claim that the genomes of a clone and the donor are the same, then we must inquire about how much difference between genomes would be necessary in order for a clone to have human dignity. Would children born from parents who are genetically related—for example, cousins—somehow have less dignity than those whose parents are not genetically related? After all, their genomes would be more similar to each other than those of unrelated people; that is, according to this view, their children would be somehow less unique than other children. And what are we to say of identical twins? Would we have to declare that they are not unique individuals? Would we maintain that they have less dignity or no dignity at all? Of course we can all see the absurdity of these claims. But if we assert that the uniqueness of our genome is what confers dignity, we must decide where the line might fall. If we discount the fact that the genome of a clone is, in fact, not identical, both because of the differences in mitochondrial DNA and because of differences in nuclear DNA, then we must decide how much difference is relevant for having dignity.

Furthermore, genomes are far from being the only relevant element in the creation of a new organism, clone or otherwise. Thus, things get even more muddled for those who argue that the uniqueness of our genome is somehow related to human dignity when we include in our discussions epigenetic changes. Epigenetic mechanisms regulate gene expression but do not involve alterations in the genetic code itself. Epigenetic modifications can abolish gene function, even without a DNA sequence change. These modifications are essential for development and differentiation, and they can also arise in mature mammals by random change or under the influence of the environment.[17] As we said before, these changes can cause heritable variations. One of these changes, for example, is the DNA-methylation system. DNA-methylation is the addition of particular chemical groups to the DNA. The number and pattern of such methylated DNA influences the functional state of the DNA material. Methyl groups block transcription of any genes to which they are attached, and when genes are not transcribed, the molecules—often proteins—for which these genes code are not produced. This methylation pattern is copied with the DNA by a special methylation system. However, changes in the pattern and number occur, hence producing different phenotypes, even when the DNA sequence has not changed. For example, our body is composed of different types of cells: muscle cells, blood cells, skin cells, and so on. We know that most cells of the mammalian body share similar DNA content. Nonetheless, despite having the same DNA, mammalian cells show a wide functional diversity. Epigenetic DNA modifications that specify which genes are expressed are responsible in part for this phenotypic diversity. Also, recent work has shown that the abnormal methylation of particular sections of regulatory genes is a substantial pathway to cancer development. Hence, cells with the same nucleotide sequence might develop in very different ways due to factors other than their DNA.[18]

Imprinting, another epigenetic change, is no less relevant for humans and other placental mammals.[19] Genomic imprinting affects a number of mammalian genes and results in the expression of some genes from only one of the two parental chromosomes. During formation of germ cells, genes subject to imprinting are marked by methylation according to whether they are present in an egg or sperm. DNA-methylation is hence used to distinguish between two copies of a gene that may be otherwise identical. Imprinting is a particularly important genetic mechanism in mammals, and it is thought to influence the transfer of nutrients to the fetus and the newborn from the mother. One example of this genomic imprinting is found in the gene that encodes for a protein called "insulin-like growth factor-2 (Igf2)."[20] This protein is required for prenatal growth, and mice that do not express it are born half the size of normal mice. However, only the paternal copy of this protein is transcribed. Hence, mice with a mutated paternally derived Igf2 gene are stunted, while mice with a defective maternally derived Igf2 gene are normal.

As we can see, epigenetic changes are therefore heritable changes in phenotypes, but they do not result from changes in the actual DNA nucleotide sequences.[21] To this, of course, we must add the effects of the uterine environment: the chemical and physical environment where the fetus grows, whether the mother gets adequate nutrients, which ones, when, and a multitude of similar factors. We must also include the effects of other developmental and environmental influences once the baby is born. Why we should ignore these possible changes and focus exclusively on the similarity of genomes in order to assign dignity is completely unclear.

Since it is scientifically inadequate to talk about the genomes of a clone and of the donor of the genetic material as identical, then to argue that cloning is contrary to human dignity on grounds that the genomes are the same would be erroneous. Critics might object that the opposition to cloning as a practice that contradicts human dignity is not grounded on the fact that the genetic material is exactly the same, but on the fact that the genome is similar enough as to produce serious psychological harms to the cloned child because of a possible loss of a sense of individuality or unique identity.[22] Thus, the focus is not so much on the genotype as it is on the phenotypic effects of the genome. Presumably, similar genotypes would produce similar enough phenotypes as to cause the clones to wonder about their own identity and individuality.

The argument, however, also seems to presuppose incorrectly that human individuality or identity is somehow determined by the uniqueness of our genome. This assumption can only be grounded on the crudest genetic determinism. As we mentioned before, according to a strong version of genetic determinism, individuals' genetic endowments completely determine who they will be. As a Nobel laureate put it, soon we will be able to pull out a CD with our own mapped genome and say, "Here is a human being; It is me."[23] Of course, these kinds of remarks can only be understood as rhetorical exaggerations. Nothing in current scientific knowledge allows us to make such statements. On the contrary, more and more evidence is accumulating that shows the complex relations between genotype and phenotypes. This is not to discount the importance of genes in human development, but only to set such importance into perspective. Hence, if it is difficult to pinpoint physical characteristics that are exclusively or even mainly determined by our DNA sequence, it is even more so to isolate genetic sequences that might be involved in the complex physiological and psychological traits that can help form a sense of individuality or personal identity.

Accordingly, the suggestion that our genome alone can be responsible for the formation of our unique identity has no scientific basis whatsoever. It is clear that whether a particular trait will be present in an individual depends not just on genes but also on other biological and environmental factors. Not surprisingly, in spite of having practically the same genes, identical twins certainly have unique and distinct personal identities. They develop different interests and relationships; they make different choices. Their individuality does not seem to be threatened by the fact that presumably they do not have unique genetic endowments. Of course, nobody has ever suggested that homozygous twins or triplets, and so on, lack human dignity. And this is significant because, as we know, their genomes are in fact much more similar than the genome of a clone and the nuclear DNA donor might be. This is so because twins share their mitochondrial DNA, while clones need not do so.

Some philosophers have argued that the psychological harm to the clone results from the violation of what Hans Jonas has called "a right to ignorance," or what Joel Feinberg has called "a right to an open future."[24] Jonas argues that human cloning, in which there is an important time gap between the beginning of the lives of the earlier and later twin, differs essentially from the simultaneous beginning of naturally occurring identical twins. According to Jonas, later twins created by cloning would know, or at least would believe they know, too much about themselves. This is so because there is already in the world another person who from the same genetic starting point has made the life choices that are still in the later twin's future. The later twin may feel that her life has already been lived, that her fate has already been determined.

Similarly, Joel Feinberg has argued that a child has a right to an open future. This, he says, requires that others raising a child do not close off the future possibilities that the child would otherwise have, as this would eliminate a reasonable range of opportunities for the child to construct his or her own life. Thus, creating a later twin could violate this right because she will believe that her future has already been set for her by the choices the earlier twin made.

Recently, Jürgen Habermas has put forward a similar argument defending the assertion that any kind of genetic manipulation is a foreclosing of a future that would otherwise be undetermined because of the natural lottery. When we design human beings either by using nuclear somatic transfer or by any other kind of prenatal genetic intervention, then, according to Habermas, we are also determining their future. As he says, "Genetically programmed persons might no longer regard themselves as the sole authors of their own life history."[25]

As in the case of the arguments about a lack of sense of individuality and uniqueness, these appeals to a right to ignorance or to an open future are questionable for several reasons. First, these arguments could not reject all cloning, but only cloning that uses nuclear DNA from an adult donor. If we wished to use the genome of a newborn baby, for example, to make a clone, problems with the right to an open future or a right to ignorance would not arise. This is so because both the baby donor of DNA and the clone would grow and develop at more or less the same time.

Moreover, these arguments rest on the disputable assumption that one's genetic endowments completely determine one's entire life path. But, as I said earlier, such an assumption has no scientific basis. It simply ignores that genotypes have a range of phenotypic expression, overlooks the importance of the environment, and disregards the significance of one's choices in building a unique and distinctive life. Nevertheless, once we reject the assumption of genetic determinism, we have no more reason to say that to produce a later clone would violate his or her right to ignorance or to an open future than we have reason to say that an older sibling would violate such rights. After all, brothers and sisters share 50 percent of their genes. And it may certainly be the case that the life choices of an older sibling influence the kinds of choices the younger one will make. If the older sister finds her career choices offer her a meaningful life, then the younger sister might decide to follow in her footsteps. If, on the other hand, the older sibling finds her life wanting because of her decisions, the younger sister may choose to follow some different path. In any case, we do not think that parents violate their younger children's right to an open future or to ignorance when they decide to bring them into the world. Similarly, parents often hold younger siblings to standards and expectations set by older sisters and brothers, and when they do follow their parent's advice, we do not believe they have violated the younger children's right to an open future.

It is true, however, that the falsity of the belief in genetic determinism only shows that a right to ignorance, or a right to an open future, is not being violated by cloning humans.[26] The falsity of this belief, nevertheless, does not show that the possible psychological harms to the clone are nonexistent, especially if the belief in genetic determinism is widespread. Two problems, however, make this argument against cloning not very compelling. First, psychological risks of this kind are presently only speculative, given that we have no experience with human cloning. Second, banning particular practices on the grounds that people's false beliefs can produce individual harms is highly controversial. The argument seems to presuppose that we have to grant decisive weight to shared false beliefs instead of, for example, trying to eliminate those beliefs by educating people. Apart from more educated people, there are additional reasons that weigh in favor of correcting people's false beliefs about genetic determinism rather than trying to accommodate them. Some of these reasons have to do with individuals' perceptions about how to respond to diseases that have a genetic component[27] or about the kinds of public policies that a community might find more adequate when dealing with health and disease.

Other critics of reproductive cloning might believe that such a procedure is contrary to human dignity because this practice can diminish our respect for human life. This is so, they claim, because cloning allows us to see human beings as replaceable: we create a clone in order to replace the DNA donor.[28] The problem with this argument is that, again, it presupposes that genes determine the individuality of persons. Only if this were the case could we say that a later clone is "replacing" another person. But, as we have said before, there is no scientific evidence to support this kind of genetic determinism. The evidence, again, points to the falsity of such an idea.

Another reason why cloning might threaten the worth of persons and as a result make cloning a practice inconsistent with human dignity is that cloning invites us to see people as made-to-order.[29] Opponents of cloning claim that people might produce children with genomes that are of special interest to those requesting the cloning. Children thus created would be valued as means and not as ends in themselves. This argument, however, is questionable because it seems to presuppose that cloning people with particular abilities or traits would guarantee that the clones would also have those same abilities or traits. For example, there is no assurance that a clone of Tiger Woods would be an exceptional golf player.

But maybe the point is not one of assurance that we will obtain another Tiger Woods, but one of probability: a clone of Tiger Woods would be more likely to become a great golf player. But, unless we incorrectly presuppose that genes determine our life paths, we have no evidence to believe such a thing. Being a golf player requires more than having particular characteristics such as having a particular height (assuming here that this trait is exclusively determined by our DNA). Tiger Woods's abilities as a golf player depend not only on his genes, but also on the environment in which he developed, his disciplined character, his drive, and, not irrelevantly, the life choices he has made. Let us remember again that identical twins can have very different abilities and can choose very different paths in their lives in spite of having similar genomes. Of course, people might have false beliefs about the role of genes, but arguments rejecting cloning because it threatens the worth of persons and therefore is contrary to human dignity only reinforce such false beliefs. Our current biological knowledge cannot support claims about genes as the only, or even the main, determinants of most human traits.

ARGUMENTS SUPPORTING CLONING

Arguments rejecting human reproductive cloning are not the only ones grounded on a misunderstanding of the role of genes in human life. Several of the arguments supporting cloning appear to suffer from this same misunderstanding. According to some scholars, "the strongest direct arguments for originating a child by nuclear somatic transfer is that his parents might give him or her a wonderful genetic legacy."[30] Hence, couples at high risk of having offspring with a genetic disease such as cystic fibrosis or Huntington's disease, for example, can decide to originate a child by cloning in order to avoid the risks of transmitting the genetic disease, while still having biologically related offspring, at least to one of the parents.

Some supporters of cloning have presented this practice as the solution to most of our deadly diseases. In the words of Gregory Pence,

> Although we do not know how much early death is caused by heritable diseases, a lot is. Until we try to prevent such deaths, we will never know how long humans can live. For the other two million Americans who die each year, the statistics aren't much better: 750,000 from heart diseases, 500,000 from cancers, and 150,000 from strokes. Over 70% of deaths may be from preventable, genetic causes. Seen in this light, originating humans by nuclear somatic transfer from healthy adults doesn't seem quite crazy.[31]

To these diseases we must also add deaths caused by other genetic diseases such as Huntington's, sickle-cell anemia, Tay-Sachs, and muscular dystrophy. Given the existence of all these preventable genetic diseases, they argue, originating children by cloning might save the lives of a considerable number of people by allowing parents to clone a child using genetic material from the nonaffected parent, or from some other "healthy" relative.

There are several problems with this argument. First, it is unclear why this situation would require cloning rather than other kinds of assisted reproductive techniques such as in vitro fertilization (IVF) with preimplantation diagnosis, or donor insemination. These technologies can help people who want to have genetically related offspring but are at risk of transmitting a serious genetic disease.

The most serious problem with this argument, however, is a misunderstanding of biology. To support cloning on grounds that it could help us prevent many of the deaths caused by heart disease, stroke, or cancer incorrectly assumes that a disease that has a genetic component is also inheritable. This is, however, erroneous. Thus, as I said before, although we can say with confidence that these diseases do have some genetic cause, we do not have evidence that they are inheritable. In fact, scientific evidence suggests that this is not so for most cases of these diseases. For example, research indicates that lung cancer, which is the leading cause of cancer death among men and women in the United States, has genetic causes.[32] These genetic causes, however, are in most cases somatic and consequently are not inheritable. Smoking, for example, produces DNA mutations of oncogenes or tumor suppressor genes in lung cells. Oncogenes are genes that promote cell division, while tumor suppressor genes are those that slow cell division or cause cells to die at appropriate times. These DNA mutations, which usually are acquired during life rather than being inherited, either activate particular oncogenes, such as the ras oncogene, or inactivate tumor suppressor genes, such as the p53 gene. The result of these acquired mutations is lung cancer. So, even if we grant that most cases of cancers and heart problems are genetic, this says nothing about whether such diseases are inheritable.

As scientific research shows, although most cases of complex diseases such as cancers, strokes, or heart disease might have genetic causes, they are not inheritable. But if this is so, arguments for reproductive cloning on grounds that this practice can help reduce the incidence of these diseases, or that parents can use it as a way to prevent these diseases in the clone, are mistaken. Moreover, even if we granted that most cases of these diseases are inheritable, for which, again, we have no scientific evidence, more than this knowledge would be needed in order to defend cloning as a way to avoid genetic diseases in our offspring. Suppose that Peter and Mary want to prevent passing on a disease-causing gene. They also need to know whether or not they have the gene. Consequently, we need to have reliable testing methods to be able to obtain this information. If these tests do not exist, then knowing that many of the diseases affecting humans are inheritable would be of little help to Peter and Mary. Of course, we might suspect that a particular disease "runs in the family." However, if it is the case that most diseases affecting humans are inheritable, then most of us will have some disease or another running in our families. Consequently, unless Peter and Mary can know with reasonable accuracy that at least one of them does not carry a particular disease-causing gene, then cloning might not prevent the passing of that gene to the clone.

It is, of course, correct to point out that there are some diseases that do have genetic causes and are inheritable. And for some of these diseases we do have genetic tests that are quite accurate, for example Huntington's disease or Tay-Sachs. However, these diseases affect a very small percentage of the human population,[33] and, more importantly, parents can avoid passing on these diseases by other less expensive and less risky means.

Another reason often given in support of human cloning is that this technique would enable some individuals to clone a person who has special meaning to them.[34] For example, parents might decide to clone their dying children or other family members. In some cases, the examples offered here are highly imaginative.[35] It is interesting to notice that this argument uses as a reason to defend reproductive cloning precisely what some of the critics of cloning have utilized as a rationale to reject this practice: the fact that the clone would look like, be similar to, and have the "same" genome as another human being. Thus, what critics find problematic, proponents of this argument seem to find appealing. If critics of cloning erred in using this as a reason against such a procedure, then supporters are also mistaken in using it as a reason for cloning human beings. A clone would not be a copy of the nuclear donor, although it would look quite a lot like the donor.

There are still other problems with supporting human cloning on this ground. First, although it might be possible that, given the chance, some people would use cloning for these purposes, there are no reasons to believe that many people would do so. Second, and more importantly, it is unclear what kind of desire we are trying to grant in these cases. There are at least two possibilities. Maybe what people who want to clone dead loved ones desire is to replace those who die with a new copy of the first person. That is, they want to have a baby who would share with the dead one some specific trait, such as strength or interest in music. Or maybe cloning that loved one is a way to accept the loss and move on with their lives.

If what supporters of this argument want is to grant the first kind of desire, then their argument is dubious because it either is grounded on a crude form of genetic determinism for which there is no scientific evidence, or it promotes the granting of desires based on false beliefs. Most who offer this argument would likely reject the idea that a clone of a loved one will be his or her replacement. They will probably agree that the new cloned child will be a different person but just with the same genes. The personality, interests, and traits of the new cloned baby might be very different. But, if supporters of this argument recognize the problems with a crude genetic determinism, and if they still defend the practice of cloning in these cases, then they are endorsing the satisfaction of desires that are based on false beliefs.

On the other hand, people might desire to clone dead loved ones as a way to move on with their lives. We should recognize that the pain of losing a loved person, especially because of a premature death, might be unbearable. Hence, attempts to palliate such suffering are laudable. However, it is unclear why encouraging human cloning is better than promoting the support of other siblings and friends, or is better than having another child by usual means.

CONCLUSION

This chapter has been concerned with what a careful attention to biological knowledge can tell us when dealing with moral and public policy debates in relation to cloning human beings. The debate over cloning human beings, like many other discussions of biotechnological issues, has often been framed by arguments that misunderstand the biology behind this practice. Tying human dignity to the uniqueness of our genome is contentious because it is unclear how a biological entity such as our genome has anything to do with human dignity. But, also, such a link is questionable because of the difficulty of deciding what exactly it means to say that two genomes are identical or the same, or whether it is even accurate to say such a thing.

Of course, if a clone's genome is identical to that of the nuclear DNA donor, still it should come as no surprise that the clone might have his or her own identity and his or her own sense of individuality. Identical twins have genomes that are more similar than the genomes of a clone and a nuclear donor need to be. Moreover, identical twins usually also share parents and social environments, while a clone from an adult individual would not. Nonetheless, twins do have their own personalities, their own characters, and their own life paths. It is certainly the case that the clone will physically look quite a lot like the donor, and it might be the case that it could make similar personal, social, or political choices. But this should be hardly surprising, much less a motive to declare cloning a practice contrary to human dignity. After all, some children look quite a lot like their parents, and many of them make choices similar to those of their parents and those of their siblings. Nonetheless, we are all well aware that it can be otherwise. We must not forget that we humans are very complex creatures influenced not just by our genes, but also by many other biological, environmental, and social factors. If, as is often said, we want to eliminate inaccurate ideas about genetic determinism, it would be appropriate to begin by paying careful attention to those arguments that inadvertently promote these inadequate views. Carefully evaluating biological knowledge can help us reject some of the arguments purporting to reject or support reproductive cloning.

It is important to point out that our current biological knowledge does not support many of the claims made by those who oppose human cloning, and therefore other arguments need to be offered to reject this procedure. However, were it the case that human dignity is diminished because individuality and identity are determined by our genomes, then, granting the fact that we find those characteristics important and that we want to preserve them, we could have good reasons to reject reproductive cloning. On the other hand, if we could offer scientific evidence that many human diseases are directly caused by genes alone, this would not be a sufficient reason to support reproductive cloning. In the next chapter, I will thus show that even if biology could sustain the arguments here presented supporting cloning, this would not be sufficient to accept this procedure. This would only be so if we presuppose not only that the biology is correct, but also that the social context in which cloning is developed and where claims about its moral adequacy are presented is irrelevant. In the next chapter, I will show the error of such an assumption.

NOTES

1. For a discussion of what it is that most people reject when they reject genetic determinism, see J. Kaplan, *The Limits and Lies of Human Genetic Research* (New York: Routledge, 2000), chap. 2.

2. See, for example, G. Pence, *Who's Afraid of Human Cloning?* (Lanham, MD: Rowman & Littlefield, 1997); L. M. Silver, *Remaking Eden: Cloning and Beyond in a Brave New World* (New York: Avon, 1997); J. Robertson, "Human Cloning and the Challenge of Regulation," *The New England Journal of Medicine* 339, no. 2 (1998): 119–22.

3. See T. Peters, *Playing God?* (New York: Routledge, 1997); U.S. National Bioethics Advisory Commission, *Cloning Human Beings: Report and Recommendations of the National Bioethics Advisory Commission* (Rockville, MD: The Commission, 1997).

4. See G. Annàs, "Scientific Discoveries and Cloning: Challenges for Public Policy," in *Flesh of My Flesh*, ed., G. Pence, 77–84 (Lanham, MD: Rowman & Littlefield, 1998); U.S. National Bioethics Advisory Commission, *Cloning Human Beings*.

5. UNESCO, *Universal Declaration on the Human Genome and Human Rights* (Paris, France: UNESCO, 1997), http://portal.unesco.org/en/ev.php-URL_ID=13177&URL_DO=DO_TOPIC&URL_SECTION=201.html (accessed 11 March 2005).

6. World Health Organization, *Ethical, Scientific and Social Implications of Cloning in Human Health* (Geneva, Switzerland: WHO, 1998), http://www.who.int/ethics/en/WHA51_10.pdf, (accessed 6 Feb. 2005).

7. Council of Europe, *Additional Protocol to the Convention for the Protection of Human Rights and Dignity of the Human Being with Regard to the Application of Biology and Medicine, on the Prohibition of Cloning Human Beings* (Paris, France: Council of Europe, 1998). Also at http://www.virtual-institute.de/en/hp/embryo/regional/AP12011998.pdf, (accessed 16 Jan. 2005).

8. The President's Council on Bioethics, *Human Cloning and Human Dignity: An Ethical Inquiry* (Washington, DC: The President's Council on Bioethics, 2002), 266, http://www.bioethics.gov/reports/cloningreport/pcbe_cloning_report.pdf (accessed 16 Jan. 2005).

9. See T. Caulfield, "Human Cloning Laws, Human Dignity and the Poverty of the Policy Making Dialogue," *BMC Medical Ethics* 4, no. 1 (2003): 3. For discussion of the significance of human dignity in bioethics, see, for example, J. D. Rendtorff, "Basic Ethical Principles in European Bioethics and Biolaw: Autonomy, Dignity, Integrity and Vulnerability—Towards a Foundation of Bioethics and Biolaw," *Medicine, Health Care, and Philosophy* 5, no. 3 (2002): 235–44; C. E. Manuel, "Physician-Assisted Suicide Permits Dignity in Dying: Oregon Takes on Attorney General Ashcroft," *The Journal of Legal Medicine* 23, no. 4 (2002): 563–86; D. Pullman, "Universalism, Particularism and the Ethics of Dignity," *Christian Bioethics* 7, no. 3 (2001): 333–58; D. Feldman, "Human Dignity as a Legal Value: Part 1," *Public Law* 4 (Winter 1999): 682–702; A. Gewirth, "Human Dignity as the Basis of Rights," in *The Constitution of Rights: Human Dignity and American Values*, ed. M. Meyer and W. Parent, 10–46 (London: Cornell University Press, 1992); O. Schachter, "Human Dignity as a Normative Concept," *American Journal of International Law* 77, no. 4 (1983): 848–54.

10. See, for example, Office of Technology Assessment, *Infertility, Medical and Social Choices* (Washington, DC: U.S. Government Printing Office, 1988); Comisión Especial de Estudio de la Fecundación "In Vitro" y la Inseminación Artificial Humanas [Special Commission for the Study of Human In Vitro Fertilization and Artificial Insemination], *Informe* [*Report*] (Madrid, Spain: Gabinete de Publicaciones, 1987); M. Warnock, *A Question of Life: The Warnock Report on Human Fertilization and Embryology* (Oxford, UK: Blackwell, 1985).

11. Evelyn Brister suggested this scenario to me.

12. B. Alberts et al., *Molecular Biology of the Cell*, 4th ed. (New York: Garland Science, 2002), chap. 14.

13. See, for example, P. F. Chinnery et al., "Accumulation of Mitochondrial DNA Mutations in Ageing, Cancer, and Mitochondrial Disease: Is There a Common Mechanism?" *Lancet* 360, no. 9342 (2002): 1323–25; S. DiMauro and E. A. Schon, "Mitochondrial DNA Mutations in Human Disease," *American Journal of Medical Genetics* 106, no. 1 (2001): 18–26.

14. Symptoms of this disease appear usually during childhood and include muscle jerking, seizures, and lack of coordination. See, for example, D. C. Wallace, "Mitochondrial DNA Mutations in Diseases of Energy Metabolism," *Journal of Bioenergetics and Biomembranes* 26, no. 3 (1994): 241–50.

15. B. Alberts et al., *Molecular Biology*, chaps. 4 and 5.

16. M. Cummings, *Human Heredity*, 5th ed. (Pacific Grove, CA: Thomson Learning, 2000), chap. 10.

17. S. Khorasanizadeh, "The Nucleosome: From Genomic Organization to Genomic Regulation," *Cell* 116, no. 2 (2004): 259–72; R. Jaenisch and A. Bird, "Epigenetic Regulation of Gene Expression: How the Genome Integrates Intrinsic and Environmental Signals," *Nature Genetics* 33, suppl. (2003): 245–54.

18. See, for example, K. P. Nephew and T. H. Huang, "Epigenetic Gene Silencing in Cancer Initiation and Progression," *Cancer Letters* 190, no. 2 (2003): 125–33; G. A. Garinis et al., "DNA Hypermethylation: When Tumor Suppressor Genes Go Silent," *Human Genetics* 111, no. 2 (2002): 115–27; P. A. Jones and P. W. Laird, "Cancer Epigenetics Comes of Age," *Nature Genetics* 21, no. 2 (1999): 163–67.

19. B. Alberts et al., *Molecular Biology*, 432–33; W. Reik and J. Walter, "Genomic Imprinting: Parental Influence on the Genome," *Nature Review Genetics* 2 (2001): 21–32.

20. B. Alberts et al., *Molecular Biology*, 432–33.

21. W. W. Gibbs, "The Unseen Genome: Beyond DNA," *Scientific American* 289, no. 6 (2003): 106–13; T. Kalebic, "Epigenetic Changes: Potential Therapeutic Targets," *Annals of the New York Academy of Sciences* 983, no. 1 (2003): 278–85.

22. See L. Kass, "The Wisdom of Repugnance," in *Flesh of My Flesh*, 13–37; U.S. National Bioethics Advisory Commission, *Cloning Human Beings*; A. D. Verhey, "Cloning: Revisiting an Old Debate," *Kennedy Institute of Ethics Journal* 4 (1994): 227–34; and D. Callahan, "Perspective on Cloning: A Threat to Individual Uniqueness," *Los Angeles Times*, November 12, 1993, B7.

23. W. Gilbert, "A Vision of the Grail," in *The Code of Codes: Scientific and Social Issues in the Human Genome Project*, ed. D. J. Kevles and L. Hood, 83–97 (Cambridge, MA: Harvard University Press, 1992).

24. H. Jonas, *Philosophical Essays: From Ancient Creed to Technological Man* (Englewood Cliffs, NJ: Prentice Hall, 1974); and J. Feinberg, "The Child's Right to an Open Future," in *Whose Child? Children's Rights, Parental Authority, and State Power*, ed. W. Aiken and H. LaFollette (Totowa, NJ: Rowman & Littlefield, 1980), 124–53.

25. J. Habermas, *The Future of Human Nature* (Cambridge, UK: Polity, 2003), 79.

26. D. W. Brock, "Cloning Human Beings: An Assessment of the Ethical Issues Pro and Con," in *Clones and Clones*, ed. M. C. Nussbaum and C. R. Sunstein (New York: W. W. Norton & Company, 1998), 153–54.

27. See chapters 7 and 8 for a discussion of this issue.

28. See U.S. National Bioethics Advisory Commission, *Cloning Human Beings*; and R. Macking, "Splitting Embryos on the Slippery Slope: Ethics and Public Policy," *Kennedy Institute of Ethics Journal* 4 (1994): 209–26.

29. See E. Shuster, "Human Cloning: Category, Dignity, and the Role of Bioethics," *Bioethics* 17, nos. 5–6 (2003): 517–25; L. Kass, "The Wisdom of Repugnance"; and U.S. National Bioethics Advisory Commission, *Cloning Human Beings*.

30. See G. Pence, *Human Cloning*, 101. See also J. Robertson, "The Question of Human Cloning," *Hastings Center Report* 24, no. 2 (1994): 6–14; and J. Robertson, "Human Cloning and Regulation," 119–22.

31. G. Pence, *Human Cloning*, 103.

32. See K. M. Fong et al., "Lung Cancer • 9: Molecular Biology of Lung Cancer: Clinical Implications," *Thorax* 58, no. 10 (2003): 892–900; H. Osada and T. Taka-

hashi, "Genetic Alterations of Multiple Tumor Suppressors and Oncogenes in the Carcinogenesis and Progression of Lung Cancer," *Oncogene* 21, no. 48 (2002): 7421–34; B. Alberts et al., *Molecular Biology*, chap. 23. For the relevance of environmental factors in cancer development, see also A. Luch, "Nature and Nurture—Lessons from Chemical Carcinogenesis," *Nature Reviews: Cancer* 5, no. 2 (2005): 113–25; L. N. Kolonel, D. Altshuler, and B. E. Henderson, "The Multiethnic Cohort Study: Exploring Genes, Lifestyle and Cancer Risk," *Nature Reviews: Cancer* 4, no. 7 (2004): 519–27; J. Peto, "Cancer Epidemiology in the Last Century and the Next Decade," *Nature* 411, no. 6835 (2001): 390–95.

33. See Institute of Medical Genetics, *Frequency of Inherited Disorders Database* (University of Wales College of Medicine) at http://archive.uwcm.ac.uk/uwcm/mg/fidd/ (accessed 21 Feb. 2005).

34. See J. Robertson, "Question of Human Cloning," 6–14; G. Pence, *Human Cloning*.

35. See G. Pence, *Human Cloning*, 60.

Chapter Five

Putting Human Cloning Where It Belongs

The main justification offered in defense of reproductive human cloning has been that this practice would alleviate human suffering. In chapter 4 we evaluated two of the ways in which such amelioration would take place: cloning could eliminate much genetic disease, and it could help those who have suffered the loss of a loved one. We also saw that both of these arguments were grounded on a misunderstanding of biological knowledge. The fact that a disease is genetic does not mean that it is inheritable. Also, our genome is not the only or main determinant of most diseases that affect human beings. Even when considering just the impact of inheritable diseases, there exist other reproductive options aside from cloning. Similarly, genomes are far from being the only or main determinants of our sense of individuality or identity, and the relevance of our genome for human dignity is far from clear. Our genes are doubtless very important, but so are many other biological, environmental, and social factors. We learned in chapter 4 what biology could tell us: that we cannot legitimately support reproductive cloning on these grounds.

This chapter focuses on what biology cannot tell us. Here I will assume that arguments defending cloning are not questionable on scientific grounds. Nevertheless, these and similar arguments err because they fail to take into account the social context in which the practice of reproductive cloning would take place. To strengthen my position, I will also discuss here another argument that has been offered in defense of cloning. This argument also has as one of its main motivations the fact that cloning would ameliorate human suffering; in this case, the pain mitigated would be that of infertile couples. After presenting and critically evaluating this new argument, I will show that if the amelioration of human suffering is our concern, a careful attention to the social and political context in which we propose to implement human cloning will show that we have good reasons not to proceed with this technology.

REPRODUCTIVE CLONING AND INFERTILITY

Those who support reproductive cloning often defend this practice by arguing that it would benefit infertile couples.[1] Human cloning would allow people who have no functional germ cells to have children genetically related to them. Estimates of the number of people who suffer from infertility vary significantly depending on the definition of infertility. Under one of the most accepted definitions—failure to conceive after twelve months of unprotected intercourse—infertility affects between 7 and 10 percent of couples with women of childbearing age.[2] Obviously, the larger the number of people who need cloning as the only means to overcome their infertility, the more acceptable cloning would appear.

Given the importance that most people attach to having children, and given the serious psychological problems that this impairment might cause to people suffering from it, concerns to relieve infertility are certainly admirable. However, maintaining that the cloning of human beings should be permitted because it can solve the problems of infertile people seems to presuppose a particular conception of infertility that is far from indisputable. First, it accepts that what is challenging about infertility is the impossibility of having genetically related children. Second, it assumes that infertility is mainly an individual medical problem in need of a technological solution, rather than also a social issue in need of social responses.

Proposing the cloning of human beings as a way to solve reproductive problems presupposes that what is problematic about infertility is the inability of having genetically related children rather than the lack of the social experience of parenting. Those who use this argument acknowledge that genetic relationships do matter in our society.[3] They often use as evidence for the importance of the genetic connection the fact that for many people the genetic bond between parents and children is the strongest possible bond. Such a connection also matters, they argue, in the extended family, given that if both parents should die, the other members of the family would feel obligated to raise the orphans. They also mention the fact that many adopted children seek to discover their biological parents, and that in many cases courts have returned adopted children to their biological parents.

As I am sure advocates of cloning using this argument know, all of these reasons acknowledging the importance of genetic relationships are descriptive claims that indicate that in our society people do value this bond. These claims do not, however, prove that prizing this genetic connection is somehow unavoidable. It is obviously the case that many human beings do not privilege these connections over other types of human relationships. In many cases, people choose to adopt over having their own children. Similarly, people do find that a network of nongenetically related friends, for example, plays a more meaningful and significant role than do genetically related family members. More importantly, from the fact that people do feel especially connected to those with whom they share genetic material, it follows neither that valuing the genetic connection is a good in itself nor that we should encourage technologies such as cloning that promote this connection.

Notice that I am not claiming that sharing a genetic link with one's offspring does not have, or should not have, any value whatsoever. It is certainly the case that, as supporters of cloning point out, many people do value genetic relationships, and that they might feel that their lives are less fulfilling without genetically related offspring. Arguably, it is also the case that the world might be a better place if all parents had the means and desire to responsibly care for and nurture their genetically related children rather than maybe putting them for adoption. Nor am I asserting that valuing a genetic relationship is a question mainly of socialization and that, therefore, people might relearn not to value this kind of tie. It might certainly be the case that as a species we have evolved to value this relationship as a strategy to assure our survival. My intention is to point out that the importance of such a connection does not, by itself, provide an incontrovertible justification for the use of technologies such as human cloning.

The understanding of infertility as mainly an individual medical problem is also questionable because the causes of reproductive difficulties and the reasons that make infertility a serious concern are, in part, socially rooted. Sexual, contraceptive, and medical practices; occupational health hazards; environmental pollution; and food additives constitute some examples of preventable causes of infertility that are socially grounded. Preventable and treatable sexually transmitted diseases (STDs) such as chlamydia, gonorrhea, and syphilis are responsible for 20 percent of the cases of infertility.[4] Thousands of women each year have to deal with procreation problems due to pelvic inflammatory disease (PID) caused by STDs.[5] Hormonal contraceptives, such as Depo-Provera, as well as others, such as intrauterine devices, increase the risk of PID and infertility. Also, according to some professionals, iatrogenic or doctor-induced infertility is common. Problems such as postchildbirth and postoperative infections can cause infertility.[6] Likewise, social practices such as delaying childbearing may be responsible for reproductive difficulties. Some evidence also suggests that environmental pollutants and chemicals can damage the reproductive capability of both women and men. Drugs such as DES can also cause infertility.[7]

Among the poor, inadequate nutrition, poor health, and limited access to health care also contribute to reproductive problems. For example, infertility is higher in poor and minority communities.[8] Black women have an infertility rate one and one-half times higher than that of white women.[9] Some of the contributing factors are a higher incidence of STDs, greater use of intrauterine devices, environmental factors (such as occupational hazards affecting reproduction), lack of access to medical treatment, nutritional deficiencies, and complications or infections following childbirth or abortion.

Similarly, it is arguable that there are social factors that make involuntary childlessness a more serious problem than it might otherwise be.[10] The problem in this case is not the existence of infertility itself, but the fact that people perceive infertility as a grave disease or handicap, and one that affects people's lives in detrimental ways. For instance, the effects of myopia in people's lives depends not only on the myopia itself, but also on social factors such as what kinds of treatment exist, how people relate to those who have myopia, whether people are ostracized when they have this problem, and so on. Some of the social elements that make the problem of infertility a very important concern are pronatalistic pressures on women to reproduce, the strong emphasis of our culture on having genetically related children, and the inextricable ties between womanhood and motherhood. Because of these expectations, infertility generates a grave problem for women.

Before we can use infertility as a justification for developing and using reproductive cloning, we need additional arguments about why we can, and should, not only value genetic relationships, but also why we should privilege them over other sorts of human relationships. Further, we need compelling arguments as to why we ought to discount the social dimensions of infertility, given that if infertility is, at least in part, socially constructed, then solutions other than high-tech ones might be more appropriate. Consequently, absent convincing arguments for the conceptualization of the problem of infertility as the individual medical problem of being unable to have genetically related children, then the argument for the cloning of humans on grounds that it will help some infertile people is not very persuasive.

CONTEXTUALIZING CLONING

Often, philosophers working in bioethics have a tendency to try to make things general, and simpler, by eliminating context. We hear about doctor-patient relationships, but in many cases these relationships are presented in a decontextualized way: no families, no communities, no institutions. We read about autonomy in ways that picture human beings as completely separated from the social environment in which they develop. We hear about the wonderful powers of genetic therapy with its ability to eliminate disease and handicaps from our lives, without considering the actual social context in which genetic technologies are implemented. Particulars such as ethnicity, economic class, and gender often seem to be lost in this ocean of generality and abstraction. However, in losing these particulars, we are neglecting the analysis of serious moral problems, such as issues of discrimination, and with it the possibility of offering some kind of solution to such problems.

The omission of social context is also present in many of the evaluations of human cloning. As we have seen, arguments for cloning often defend the claim that this technology will reduce human suffering in a number of ways. They maintain that it will help the infertile, that it will reduce genetic disease, and that it will help people who suffer because of the loss of loved ones. When these forms of suffering are considered outside of the broader social context of human good and affliction, they seem to grant the need for cloning. However, when they are placed within our current context of limited resources and the existence of many other diseases, disabilities, and impairments, we realize that if our concern is to reduce human misery, then cloning cannot be justified.

In a world with limited resources and where numerous diseases and impairments affect people, one must ask what is it about infertility that attracts so much attention. Certainly, the inability to conceive can be stressful, and painful, but this is the case with many other diseases that have not drawn such interest. If we lived in a world where most major causes of stress and pain (even if only medical ones) were solved, one might think that if a new technology appears that can solve infertility, its development should be supported. But, unfortunately, we are far from living in such a world. Many of the diseases and impairments that humans endure are life threatening, and some of them are not. In 2002, for example, the overall percentage of Americans living in poverty was 12.1 percent; the rate of low birth weight, which is associated with elevated risk of death and disability in infants, was 7.8 percent.[11] Furthermore, 11.2 percent of noninstitutionalized adults were diagnosed with heart disease, and nearly 700,000 were killed by it.[12] Seven percent of adults were diagnosed with cancer, which killed over 550,000.[13] Diabetes was diagnosed in 13.4 million people, and over 73,000 died of the condition.[14] And these are data only from the United States. If we take into account global trends in mortality and morbidity, the situation looks much more dire. It is obvious that in this situation priorities need to be set, and good reasons must be offered for those priorities. Infertility, though a source of suffering, is not a life-threatening condition.

Also, the argument that the cloning of humans is permissible because it might benefit the infertile could have more strength if the number of people who would be helped by this technique were significant. Arguably, solutions for diseases that cause a significant amount of suffering and distress, even when not life threatening, and affect a significant amount of the population, would have priority over similar disorders that affect a small percentage of people. However, as even proponents of this argument for reproductive cloning recognize, there are no reasons to believe that the number of people using this technology would be large, especially if it is considered a technology of last resort.[15] Thus, unless we want to argue that the good of relieving the suffering of those who cannot have their own children, and who have access to expensive and risky technologies such as cloning, outweighs the good of justly allocating scarce medical resources, then the development and use of cloning as a way to reduce suffering is not morally warranted.

Some have argued that there is an important group of people, gays and lesbians, who might also be benefited by human cloning techniques.[16] This group ought to be considered because helping them to relieve infertility is good not only because it alleviates their suffering, but also, and more importantly, because assisting members of a discriminated group is a healthy sign of solidarity. I think this argument stresses a significant point, the importance of solidarity. However, the argument fails for several reasons. First, it emphasizes the idea that infertility is mainly a problem resulting from the inability to have genetically related children. Second, to diagnose homosexuals as infertile (even if only temporarily) presents an aspect of homosexuality as a medical condition in need of medical treatment.[17] Furthermore, there are certainly other means by which homosexuals can become parents, such as adoption or artificial insemination. Also, and more importantly, allowing homosexuals to have children by using cloning techniques is unlikely to solve issues of unfair treatment. It is also unlikely to help the many homosexual couples that might be unable to afford this technique. If the problem is the discrimination that homosexuals face in different aspects of their lives, then we need to change the social environment and institutions that promote or sustain such discrimination. Problems of discrimination are rarely solved with technological fixes.

Furthermore, if our concern is to solve infertility, it is certainly the case that other means to relieve the problem might be more effective.[18] But, when proponents of cloning frame the issue as an individual medical problem in need of a technological solution, they might be undermining their own attempts to fight infertility. This is so because, in emphasizing technical responses to reproductive problems, supporters of cloning draw attention away from the fact that, as we have seen earlier, many of the causes of infertility could be prevented. Thus, stressing the importance of developing cloning as a solution to infertility might promote public policies that would result in funds being dedicated mainly to high-technological solutions rather than also to preventive measures. These measures could include stricter controls for environmental pollutants and chemicals, more research funding for safer contraceptives, and educational programs to prevent STDs and for treating them before they cause reproductive difficulties. Similarly, if we ignore the fact that infertility is in part a socially created problem, then proponents of cloning have also missed the opportunity to see other solutions to the problem. Some such solutions could be implementing social policies that could help to modify the view of motherhood as the primary role of women, encouraging an understanding of maternity as a possible but not as a necessary choice, facilitating adoption, or promoting different forms of mothering. All of these possible solutions would certainly help to relieve the suffering of a much greater number of people. Thus, it is reasonable to believe that in a more just society, limited resources would be better spent also on social rather than mainly on technological solutions to infertility.

Someone might object that those who support cloning because it might help infertile people might also defend other ways to solve the problem. What is significant, however, is that although many proponents of the infertility argument spend considerable time reflecting on the problem of infertility, they rarely acknowledge nontechnological solutions as a way to solve reproductive difficulties.[19] In fact, I am arguing that advocating these kinds of social solutions would undercut the need for a technological fix such as the cloning of human beings. If people at risk of suffering STDs have access to education and medical care, this would decrease the incidence of these diseases, and with it, the incidence of infertility. The same can be said about other preventable causes of infertility such as lack of nutrition, environmental hazards, iatrogenic sources, and so on. Moreover, social policies and institutional changes directed toward eliminating the image of women as essentially mothers or toward removing barriers that force women to delay childbirth if they want to have a career could be very effective in diminishing both the incidence of infertility and the suffering caused by it.

Proponents of human cloning as a way to solve infertility might argue that they are well aware of the social context in which this technology is being developed and may be implemented. This is a context of economic capitalism, limited health care access, lack of political and public desire or interest in offering affordable health care, and skyrocketing insurance premiums. It is precisely because of this awareness that they are disinclined to defend other possible solutions to the problem of infertility. Furthermore, allocating funds for prevention may seem less reasonable than financing treatments such as cloning because the latter meets the needs of identifiable individuals. Preventive programs also help people, but it is difficult to identify them. People sympathize with someone anxiously awaiting a medical intervention such as cloning to have a child that they cannot have by other means. It is more difficult to generate empathy for the unidentifiable individual whom prevention programs may help, even if the suffering avoided is equally great. Hence, proponents of cloning might object to my analysis by arguing that solutions to infertility of the kind I propose here would require unattainable institutional changes.[20] Certainly, some alternatives to the problem of infertility would likely require social transformations such as changes in attitudes toward women and motherhood or alterations in family structures. Prevention would also involve educative programs, social services, legislative changes in occupational health and safety, and environmental legislation. These changes are difficult because of economic pressures and the extensive time period required to obtain results.

Although this criticism raises important concerns, it is incorrect for several reasons.[21] First, if we discard policy options before we adequately evaluate them, it is difficult to see how we can talk about "unattainable" institutional changes. To affirm that alternatives to the problem of infertility such as prevention require unworkable changes without giving some evidence for such an argument is, then, dubious. Moreover, if we only take into account policy options that are highly feasible under the technological and economic status quo, then our evaluations erroneously encourage a self-fulfilling prophecy sanctioning current conditions, regardless of their worth. Furthermore, development of cloning would require significant amounts of public and private funds, and each instance of use of the technology would be quite costly. On the other hand, reduction in occupational and environmental hazards would affect not just the incidence of infertility but would also help with other aspects of human health.

While infertility is not a life-threatening condition, some genetic diseases lead to premature deaths or seriously impaired lives. Thus, a defense of reproductive human cloning in this case might appear better founded. Nevertheless, when we contextualize this issue, we find once again that whether reproductive cloning is an appropriate response to human suffering depends, not only on biological considerations, but also on an understanding of the social context where this technology would be used. Consequently, if other techniques exist that can help us avoid hereditary diseases, those who support cloning on this ground need to prove not only that this technology is good, but also that it is better than other techniques. At present, there are certainly other options that can be used to avoid the risk of transmitting particular genetic diseases. Parents can decide not to reproduce if they are at risk of transmitting a serious genetic disease; they can opt for adoption; they can choose to use sperm or egg donors (although these methods might be unappealing to those who value a genetic connection to their children); also available to them is preimplantation genetic diagnosis (PGD) of embryos created by IVF. Consider again, for example, Peter and Mary. Suppose that they are both carriers of a serious genetic disorder and have a 25 percent chance of having an affected child. If they use the genetic material of, say, one of Peter's cells to clone a child, then they can be reasonably sure that, although the child would also be a carrier, it would not suffer the disease. But why, we must ask, would Peter and Mary choose cloning over IVF with PGD? As we saw in chapter 3, cloning, like IVF, requires the recovery of eggs from Mary and the transfer of the reconstructed embryo or embryos into her body in order to obtain a pregnancy. Certainly, IVF and PGD are expensive and not very successful procedures, but at present there is no reason to believe that cloning would be cheaper than other reproductive techniques. Moreover, we have evidence that shows that it would be less successful and certainly more risky for mothers and their children.[22] As we saw in chapter 3, several studies have shown that for every clone that arrives to adulthood, there are dozens that do not survive or that suffer serious health problems. It is true that in the future the technology might be perfected, and thus these health concerns might disappear. At present, however, the evidence seems to pinpoint difficulties in solving possible health problems.[23]

It is also unclear that using cloning to prevent genetic disease will help a large number of people. As we saw in chapter 4, the cases in which the existence of a particular gene guarantees that a particular disease will appear are rare. Similarly, even if we can agree that many diseases affecting human beings have a genetic component, this does not mean that they are inheritable. Thus, reproductive cloning would be of little help in those cases.

Of course this should not be understood as an argument that we should not attempt to develop technologies that might help a small number of people. Certainly, it may be justifiable to use resources to try to cure very rare cancers, for instance, or other rare diseases. The use of reproductive cloning to prevent inheritable diseases, however, would not address any actually existing suffering. Cloning, in this case, is used in order to avoid the birth of an affected infant. Those who are at risk of transmitting genetic diseases already have other means, maybe even better means, to avoid doing so. Reproductive cloning is not intended to save anybody's life. It is not intended to cure existing people.

No less important is the fact that, in our world, preventing most cases of premature death requires not the elimination of genetic disease, but access to simple vaccinations such as those for measles, tuberculosis, or influenza. It requires access to basic amounts of food and clean water so as to avoid malnutrition, dehydration, and a multitude of infectious diseases. It requires access to preventive health care for mothers and their children so as to make more likely healthy newborns, robust mothers, and thriving children. It requires the development of social structures that promote, for example, the prevention of traffic accidents, especially among teenagers, or the elimination of environmental toxins such as lead, asbestos, and PCBs. Any of these actions would ameliorate human suffering more efficiently than would reproductive human cloning.

Similarly, although it might be possible that, given the chance, some people would use cloning to replace people who have special meaning to them, there are no reasons to believe that many would do so. And, again, given a world with limited resources and with other more pressing needs, to support the development and use of a quite likely very expensive, not very efficient, and quite risky technology to help a small number of people seems questionable.

When one reads analyses of human cloning, one has the impression that we live in a society where the most serious and pressing problems are the pleas of infertile people, or the requests of those who want to replace their dead loved ones, a world where genetic disease is the main cause of preventable deaths, and where resources are all but unlimited. And, probably, in a world where these were people's main sources of suffering, the kind of debate about human cloning that is occurring now would make perfect sense. But, that is not the world we live in. Ours is an overpopulated world where thousands of children are in desperate need of good homes, a world where thousands of mothers who are lucky enough to have children of their own lack access to basic health care for their children or are unable to provide nutritious food or safe water for them. In the United States, forty-five million people lacked health care insurance in 2003.[24] Over 11 percent of American children are without health insurance, and nearly 6 percent are without a usual source of health care.[25] In 2003, 12.5 percent of people in the United States, amounting to nearly thirty-six million Americans, were below the poverty line. Of those, 17.6 percent were children under eighteen years of age.[26] When we set our discussion of human cloning in this world of ours—that is, when we do not lose sight of the social and political context in which this technology might be developed and implemented—it seems that deciding how cloning might be legitimately used to relieve the pain of those who cannot have their own children, or of those who request human cloning as a way to have genetically related children without genetic diseases, or of those who solicit this new technology in order to cope with the pain of losing a loved one is not our most pressing moral and public policy concern. Again, if, as the arguments offered to support human cloning clearly suggest, the main goal of developing and using this technology is to reduce human suffering, and unless they want to argue that the suffering of those who endure infertility, genetic diseases, or the loss of loved ones and who are lucky enough to be able to afford this technology trumps consideration about the affliction of other people in need, then they have to agree that human cloning is far from being our best response to human misery.

Moreover, we should not neglect the fact that reproductive cloning is a risky procedure. Therefore, not only might it not be the best way to alleviate human suffering, it might actually increase it. As we saw in chapter 3, present cloning technologies result in a high rate of birth defects such as respiratory disorders, fetal growth defects, and placental abnormalities, each of which can be associated with a high rate of neonatal mortality. As several studies with different species have documented to date, only 1 to 3 percent of embryos produced by cloning techniques survive to adulthood. And those who survive into adulthood might also suffer serious health problems.[27] Furthermore, this procedure is also risky for women's health and life because of placental abnormalities, increased size of fetuses, and spontaneous abortions late in pregnancy. Thus, the motivation to clone humans in order to help infertile couples, to relieve the suffering of those who have lost loved ones, or to help people at risk of transmitting a genetic disease is misguided.

I realize that some proponents of human cloning might see my arguments as controversial attempts to change the world, and that they would prefer for human cloning to be evaluated in its own sphere.[28] This criticism is, however, seriously problematic because it seems to presuppose that a decontextualized evaluation of human cloning is an adequate one. My point here has been to argue that such is not the case. Also, when we try to analyze human cloning "in its own sphere," we are implicitly and uncritically sanctioning the status quo, and thus we fail to contribute to the transformation of unjust social structures and public policies.

CONCLUSION

This chapter has focused on what biology cannot tell us when dealing with the morality of human cloning. Here I have assumed that a defense of human cloning based on the belief that this practice would help reduce human suffering is not dubious on scientific grounds. In spite of that, these and similar arguments are unsuccessful when we take into account the social context in which the practice of reproductive cloning would take place. We have seen here that if the amelioration of human suffering is our concern, careful attention to the social and political context in which we propose to implement human cloning shows that we have good reasons not to proceed with this technology.

Let me emphasize that the arguments I have presented in this chapter should not be understood as disregarding the importance of the pleas addressed by proponents of cloning humans. On the contrary, I believe that a contextualized analysis of this technology might indicate better ways to bring relief to people suffering from infertility or from genetic diseases, and to those coping with the death of loved ones. Neither am I maintaining that we should put an end to any new technologies until more basic problems are solved. The United States is a pluralistic society with competing interests that need to be considered. This is only an argument to not lose sight of the social context in which our new technologies appear. It is an argument calling attention to the fact that adequate discussion of the morality of cloning, or of any other new technology for that matter, requires not only discussion of risks and benefits—that is, a discussion of means—but also, and more importantly, a discussion of ends.

NOTES

1. See, for example, J. Robertson, "Human Cloning and the Challenge of Regulation," *The New England Journal of Medicine* 339, no. 2 (1998): 119–22; G. Pence, *Who's Afraid of Human Cloning?* (Lanham, MD: Rowman & Littlefield, 1997), 106–8; L. M. Silver, *Remaking Eden: Cloning and Beyond in a Brave New World* (New York: Avon, 1997); R. Winston, "The Promise of Cloning for Human Medicine," *British Medical Journal* 314, no. 7085 (1997): 913–14.

2. See New York Task Force on Life and the Law, *Assisted Reproductive Technologies: Analysis and Recommendations for Public Policy* (New York: The Task Force, April 1998), 10–16.

3. See, for example, G. Pence, *Human Cloning*, 108–12. See also R. Dawkins, *The Selfish Gene*, 2nd ed. (New York: Oxford University Press, 1989).

4. See G. B. Ellis, "Infertility and the Role of the Federal Government," in *Beyond Baby M*, ed. D. M. Bartels et al. (Clifton, NJ: Humana Press, 1990), 111–30.

5. See R. Jewelewicz and E. E. Wallach, "Evaluation of the Infertile Couple," in *Reproductive Medicine and Surgery*, ed. E. E. Wallach and H. A. Zacur, 364 (St. Louis, MO: Mosby, 1994); and B. A. Mueller and J. R. Daling, "The Epidemiology of Infertility," in *Controversies in Reproductive Endocrinology and Infertility*, ed. M. R. Soules, 1–13 (New York: Elsevier, 1989).

6. See R. Rowland, *Living Laboratories* (Bloomington, IN: Indiana University Press, 1992), 231, 257.

7. See R. Jewelewicz and E. E. Wallach, "Evaluation of the Infertile Couple," 364; R. Rowland, *Living Laboratories*, 231; and R. Koval and J. A. Scutt, "Genetic and Reproductive Engineering—All for the Infertile?" in *Baby Machine*, ed. J. A. Scutt, 33–57 (Melbourne, Australia: McCulloch Publishing, 1988).

8. See New York Task Force on Life and the Law, *Assisted Reproductive Technologies*; M. S. Henifin, "New Reproductive Technologies: Equity and Access to Reproductive Health," *Journal of Social Issues* 49, no. 2 (1993): 61–74.

9. See L. Nsiah-Jefferson, "Reproductive Laws, Women of Color, and Low-Income Women," in *Reproductive Laws for the 1990s*, ed. S. Cohen and N. Taub, 23–67 (Clifton, NJ: Humana Press, 1989).

10. See, for example, R. Arditti, R. D. Klein, and S. Minden, *Test-Tube Women* (London: Pandora Press, 1984); B. K. Rothman, *Recreating Motherhood* (New York: W. W. Norton & Company, 1990); A. Phoenix, A. Woollett, and E. Lloyd, eds., *Motherhood* (London: Sage, 1991); M. S. Ireland, *Reconceiving Women* (New York: Guilford Press, 1993); and R. Jackson, *Mothers Who Leave* (London, UK: Pandora, 1994).

11. National Center for Health Statistics, *Health, United States, 2004* (Hyattsville, MD: U.S. Government Printing Office, 2004), 7–8. Also available at http://www.cdc .gov/nchs/data/hus/hus04.pdf (accessed 14 Jan. 2005).

12. National Center for Health Statistics, *Fastats: Heart Disease* (Hyattsville, MD: U.S. Government Printing Office, 2004), http://www.cdc.gov/nchs/fastats/heart.htm (accessed 21 Feb. 2005).

13. National Center for Health Statistics, *Fastats: Cancer* (Hyattsville, MD: U.S. Government Printing Office, 2004), http://www.cdc.gov/nchs/fastats/cancer.htm (accessed 21 Feb. 2005).

14. National Center for Health Statistics, *Fastats: Diabetes* (Hyattsville, MD: U.S. Government Printing Office, 2004), http://www.cdc.gov/nchs/fastats/diabetes.htm (accessed 21 Feb. 2005).

15. See G. Pence, *Human Cloning*, 145.

16. See, for example, P. Kitcher, "Whose Self Is It, Anyway?" *The Sciences* 37, no. 5 (1997): 58–62.

17. For a compelling view on some of the possible problems resulting from diagnosing lesbians as infertile, see J. S. Murphy, "Should Lesbians Count as Infertile Couples? Antilesbian Discrimination in Assisted Reproduction," in *Embodying Bioethics*, ed. A. Donchin and L. Purdy, 103–20 (Lanham, MD: Rowman & Littlefield, 1999).

18. See I. de Melo-Martín, *Making Babies: Biomedical Technologies, Reproductive Ethics, and Public Policy* (Dordrecht, the Netherlands: Kluwer, 1998), chap. 5.

19. See J. Robertson, "Human Cloning and Regulation"; G. Pence, *Human Cloning*.

20. See G. Pence, *Human Cloning*, 144.

21. See I. de Melo-Martín, *Making Babies*, chap. 5.

22. For difficulties establishing pregnancies by nuclear somatic transfer, see, for example, R. S. Lee et al., "Cloned Cattle Fetuses with the Same Nuclear Genetics Are More Variable Than Contemporary Half-Siblings Resulting from Artificial Insemination and Exhibit Fetal and Placental Growth Deregulation Even in the First Trimester," *Biology of Reproduction* 70, no. 1 (2004): 1–11; M. J. Sansinena et al., "Production of Nuclear Transfer Llama (Lama Glama) Embryos from In Vitro Matured Llama Oocytes," *Cloning Stem Cells* 5, no. 3 (2003): 191–98; R. Jaenisch et al., "Nuclear Cloning, Stem Cells, and Genomic Reprogramming," *Cloning Stem Cells* 4, no. 4 (2002): 389–96; D. Solter, "Mammalian Cloning: Advances and Limitations,"

Nature Review: Genetics 1, no. 3 (2000): 199–207; I. Wilmut et al., "Viable Offspring Derived From Fetal and Adult Mammalian Cells," *Nature* 385 (1997): 810–13.

23. See, for example, R. K. Ng and J. B. Gurdon, "Epigenetic Memory of Active Gene Transcription Is Inherited Through Somatic Cell Nuclear Transfer," *Proceedings of the National Academy of Sciences USA* 102, no. 6 (2005): 1957–62; J. P. Renard et al., "Nuclear Transfer Technologies: Between Successes and Doubts," *Theriogenology* 57, no. 1 (2002): 203–22; K. Illmensee, "Cloning in Reproductive Medicine," *Journal of Assisted Reproduction and Genetics* 18, no. 8 (2001): 451–67.

24. C. DeNavas-Walt, B. D. Proctor, and R. J. Mills, U.S. Census Bureau, Current Population Reports, P60-226, *Income, Poverty, and Health Insurance Coverage in the United States: 2003* (Washington, DC: U.S. Government Printing Office, 2004), 14. Also at http://www.census.gov/prod/2004pubs/p60-226.pdf (accessed 21 Feb. 2005).

25. National Center for Health Statistics, *Health, United States, 2004*, 256; and C. DeNavas-Walt, B. D. Proctor, and R. J. Mills, *Income*, 14.

26. C. DeNavas-Walt, B. D. Proctor, and R. J. Mills, *Income*, 9.

27. See, for example, R. Yanagimachi, "Cloning: Experience from the Mouse and Other Animals," *Molecular and Cellular Endocrinology* 187, nos. 1–2 (2002): 241–48; R. Mollard, M. Denham, and A. Trounson, "Technical Advances and Pitfalls on the Way to Human Cloning," *Differentiation* 70, no. 1 (2002): 1–9. See also chapter 3 for a discussion of this issue.

28. See G. Pence, *Human Cloning*, 144.

Chapter Six

Obtaining Genetic Information

In 1953, James Watson and Francis Crick published in *Nature* the three-dimensional molecular structure of DNA, presenting what would be a breakthrough discovery in the biological sciences.[1] The Watson-Crick model of DNA resulted in remarkable theoretical and technological achievements during the next decades. The genetic code was deciphered; the cellular components, as well as the biochemical pathways, involved in DNA replication, translation, and protein synthesis were carefully described, and the enzymes responsible for catalyzing these processes were isolated.[2]

A striking result of these theoretical advances was the ability to use a variety of techniques that would allow researchers to control and manipulate DNA.[3] The discovery of restriction enzymes was one of the most important steps in this ability to manipulate DNA material. These enzymes are bacterial proteins that can recognize and cleave specific DNA sequences. They act as a kind of immune system, protecting the cell from the invasion of foreign DNA by acting as chemical knives or scissors. The capacity to cut DNA into distinct fragments was a revolutionary advance. For the first time, scientists could segment the DNA that composed a genome into fragments that were small enough to handle. Additionally, methods for synthesizing DNA and for using messenger RNA to make DNA copies provided reliable means for obtaining DNA. Moreover, they now had the opportunity to separate an organism's genes, remove its DNA, rearrange the cut pieces, or add sections from other parts of the DNA or from other organisms. The use of plasmids, extrachromosomal genetic elements found in a variety of bacterial species, and of bacterial viruses as vectors, or vehicles, to introduce foreign DNA material into living cells served as a major tool in genetic engineering. Once introduced into the nucleus, the foreign DNA material is inserted, usually at a random site, into the organism's chromosomes by intracellular enzymes. In some rare occasions, however, a foreign DNA molecule carrying a mutated gene is able to replace one of the two copies of the organism's normal gene. These rare events can be used to alter or inactivate genes of interest. This process can be done with stem cells, which can eventually give rise to a new organism with a defective or missing gene, or with somatic cells in order to compensate for a nonfunctioning gene.

No less important for the ability to understand and manipulate genetic material were the development of techniques to sequence DNA, the establishment of the methodology for gene cloning, and the development of the polymerase chain reaction (PCR). With these techniques it was possible to obtain and analyze unlimited amounts of DNA and RNA within a short period of time. Additionally, PCR would prove an invaluable method to identify mutations associated with genetic disease; to detect the presence of unwanted genetic material, for example, in cases of bacterial or viral infection; and to use in forensic science. Researchers working on organisms such as worms developed technologies that allowed mapping of their genomes. These mapping techniques permitted the location of the positions of known landmarks throughout the organism's chromosomes. Furthermore, as these molecular techniques improved, their application to cancer studies became more and more common, leading to the discovery of viruses that were able to transform normal cells into cancer cells, and to the description of oncogenes, cancer suppressor genes, and a variety of other molecules and biochemical pathways involved in the development of cancer.

The automation of DNA sequencing in the 1980s brought to the forefront of the scientific community the possibility of not just mapping the human genome, but also sequencing it. Thus, while gene mapping allowed researchers to determine the relative position of genes on a DNA molecule and the distance between them, sequencing let them identify one by one the order of bases along each chromosome.[4] The launching, in 1990, of the Human Genome Project, which culminated in the publication of the full sequence of the human genome in April 2003,[5] also brought about the development of a variety of what are now common biotechnologies. Genetic tests and screening for several human diseases such as Tay-Sachs, sickle-cell anemia, Huntington's disease, and breast cancer are now part of medical practice. Agricultural products such as cotton or soybean plants genetically modified to produce selective insecticides or tomatoes engineered to delay ripening are not unusual products in our shopping bags. Animal cloning does not make the front page anymore because it is becoming routine. Genetic therapy and pharmacogenetics are more and more often presented as the new medical miracles. And, of course, discussions of genetic enhancement and the hopeful, or frightening, possibility of designer babies are regular visitors to the news and entertainment media.

GENETIC TESTING

The term "genetic technologies" usually covers a variety of techniques. It can refer to relatively simple automated processes that test for a particular genetic mutation, as in genetic testing. It can denote complex medical procedures that attempt to introduce genetic material into somatic or germ cells to compensate for abnormal genes and make a necessary protein, as in gene therapy. And it can also refer to technologies that are directed toward altering a functional gene so as to enhance a particular organism, as in genetic enhancement. Because the main concern of the following chapters is with issues related to obtaining genetic information about our family members or ourselves, this chapter will focus on genetic testing.

Like any other kind of medical test, genetic testing gives information about someone's health status. It involves examining a person's chromosomes, DNA, RNA, or proteins to diagnose or rule out particular genetic disorders or to predict the likelihood of suffering such disorders in the future. The tests usually are performed on a sample of blood, but other tissues such as amniotic fluid, embryonic cells, hair, saliva, or skin can also be used. Once the tissue sample is obtained, laboratory technicians look for specific changes in chromosomes, DNA, RNA, or particular proteins that might indicate the existence of a disorder or the possibility of developing it. Although genetic testing was initially used to identify rare inherited disorders that usually affected a very small percentage of the population—for example, Tay-Sachs, Huntington's disease, or cystic fibrosis—as more and more tests appear, this is changing. Now tests exist, and many are being developed, that detect genetic alterations that may influence more complex and more common conditions, for example, breast and ovarian cancer, cardiovascular diseases, colon cancer, or Alzheimer's disease. Currently, genetic tests are used in a variety of situations[6]:

a. Preimplantation genetic diagnosis. This test is used in conjunction with in vitro fertilization. The embryos thus produced are then tested to identify genetic abnormalities. It is usually offered to couples with an increased risk of having a baby with a genetic or chromosomal disorder.
b. Prenatal diagnosis. It is used to detect genetic or chromosomal abnormalities, such as Down syndrome or Tay-Sachs disease, in developing fetuses. As in the case of preimplantation diagnosis, this test is offered to couples

when there is suspicion that the child might be at an increased risk of having a genetic disorder.

c. Newborn screening. It is the most common type of genetic testing. It is
carried out in newborn babies, normally as part of public health programs,
in order to identify certain diseases for which early diagnoses and treatment exist. In the United States, for example, all states currently test infants for phenylketonuria and hyperthyroidism.

d. Carrier testing. The purpose of this test is to establish whether an individual carries a copy of an altered gene that, when present in two copies,
might cause a genetic disease. Often this test is offered in the context of
reproductive planning in order to determine the risks of having a child
with a particular genetic disorder.

e. Diagnostic testing. This test attempts to confirm whether an individual has
a genetic or chromosomal condition. It is usually offered to people who
show some signs of suffering a particular disease or who have a family
history of the disease. Diagnostic tests can be performed at any time during a person's life span.

f. Predictive testing. It is used to identify mutations that might increase an
individual's risk of developing a disorder with a genetic basis later in life.
Healthy people, with or without a family history of a particular disease,
can be candidates for this type of test.

g. DNA fingerprinting. This type of test is not employed to detect genes or
mutations that are associated with particular disorders. As with fingerprinting, this test is used to identify particular individuals, usually for legal reasons, such as to implicate or exclude a crime suspect or to establish
paternity.

What all these test types have in common is that they give us information
about our genetic endowment or that of our offspring. These tests can thus be
helpful in making reproductive decisions and in health care planning. The
usefulness of the tests, however, varies depending on the condition that they
try to determine. That is, those tests that have a high predictive power, such
as, for example, tests to identify genetic mutations that are involved in Huntington's disease, are obviously more helpful than those with a low predictive
power, such as tests that are used to detect mutations involved in common and
complex diseases such as cancer.

As with reproductive cloning earlier, this chapter's purpose is to offer a
brief introduction to genetic technologies so that we can evaluate the moral
issues resulting from the use of these techniques. As we will see in the following chapters, the novel ability to obtain information about our genetic
makeup has raised new ethical dilemmas. Some scholars have argued that this

new capability results in moral obligations to ourselves and others. In chapter 7, we will see that biological knowledge cannot ground such obligations, while chapter 8 will show that other issues related to the context in which these new genetic technologies are used also call into question the existence of these presumed moral duties.

NOTES

1. J. Watson and F. Crick, "A Structure for Deoxyribose Nucleic Acid," *Nature* 171 (1953): 737–38.

2. See, for example, R. B. Macgregor and G. M. Poon, "The DNA Double Helix Fifty Years On," *Computational Biology and Chemistry* 27, nos. 4–5 (2003): 461–67; C. P. Lorentz et al., "Primer on Medical Genomics Part I: History of Genetics and Sequencing of the Human Genome," *Mayo Clinic Proceedings* 77, no. 8 (2002): 773–82; A. Ahmadian and J. Lundeberg, "A Brief History of Genetic Variation Analysis," *Biotechniques* 32, no. 5 (2002): 1122–24, 1126, 1128; D. Kevles and L. Hood, *The Code of Codes* (Cambridge, MA: Harvard University Press, 1992); J. Watson et al., *Recombinant DNA*, 2nd ed. (New York: W. H. Freeman, 1992); S. Wright, "Recombinant DNA Technology and Its Social Transformation 1972–1982," *Osiris* 2 (1986): 303–60.

3. See note 2 for references.

4. See note 2 for references. See also J. Weissenbach, "The Human Genome Project: From Mapping to Sequencing," *Clinical Chemistry and Laboratory Medicine* 36, no. 8 (1998): 511–14.

5. See U.S. Department of Energy, Office of Science, Human Genome Project Information, *About the Human Genome Project*, http://www.ornl.gov/sci/techresources/ Human_Genome/project/about.shtml (accessed 16 Feb. 2005).

6. See, for example, U.S. Department of Health and Human Services, *Understanding Genetic Testing*, http://www.accessexcellence.org/AE/AEPC/NIH/ (accessed 21 Feb. 2005); U.S. Department of Energy, Human Genome Project Information, *Gene Testing*, http://www.ornl.gov/sci/techresources/Human_Genome/medicine/genetest .shtml (accessed 21 Feb. 2005); Genetics and Public Policy Center, *Genetic Testing* (Genetics and Public Policy Center, 2005), http://www.dnapolicy.org/genetics/testing .jhtml (accessed 21 Feb. 2005).

Chapter Seven

Genetic Information and
Moral Obligations

This is the era of genetics. The news and popular press are teeming with information about genetic science, genetic technologies, genetically based diseases, and animal and human cloning. Together with the revolution in information technologies, and sometimes going hand in hand with them, the biotech revolution promises to transform our world. The well-known successes of molecular biology in the 1950s and 1960s—the double helix model of DNA structure, the operon model of gene regulation, and the genetic code—have transformed biology and especially genetics. Scientists in this discipline promise to provide a complete catalog of human genes. Every week, researchers report on the links between one or more genes and diseases such as cystic fibrosis, Tay-Sachs, heart diseases, and various types of cancer. They also inform us of new genes that appear to be determinant of health disorders, such as alcoholism, depression, or schizophrenia, which we would like to control or eradicate. The Human Genome Project promises to offer an immense array of benefits by giving us information about the diagnosis, treatment, and prevention of many, if not most, human diseases. It appears to have the ability to alter our understanding of who we are, to change the kind of medicine we practice, and to increase our capacity to control our own destiny.[1]

Scientists, the medical profession, philosophers, social scientists, policy makers, and the public at large have been quick to embrace the accomplishments of genetic science. The mapping of the human genome was presented as the first step, on a list of apparently few such steps, in solving the medical problems that afflict human beings.[2] Presentations of genetic therapy as the new revolution in medicine,[3] concerns about cloning human beings,[4] debates about the rightness or wrongness of genetic enhancement,[5] questions about enacting public policies that would prevent the use of genetic information to discriminate against people,[6] and debates about what moral obligations follow from our newly found ability to obtain genetic knowledge about ourselves and others[7] all seem to presuppose that the new science of genetics has the power to greatly transform our lives and our world. Thus, for example, it would make little sense to ponder the correctness or incorrectness of enhancing human traits unless one assumes that tinkering with the human genome is sufficient, even if not necessary, to "enhance" our memory, intelligence, disease resistance, or beauty. Similarly, it would seem odd to rush into requirements to pass legislation preventing insurance companies or employers from using genetic information to discriminate against people when we have no such legislation preventing such use of other types of medical information, unless we believe that genetic information is somehow more deterministic of future states than are other kinds of medical data, such as cholesterol levels, blood pressure measures, or levels of activity. Likewise, as we have seen in previous chapters, the flood of arguments against human cloning on grounds of a possible loss of a sense of individuality or unique identity presupposes that our genes determine human individuality or identity. Also, debates about whether we have an obligation to gain information about our genetic makeup, a duty not to bring affected children into the world, or an obligation to inform family members about genetic risks are meaningless unless we are able to not just perform tests that will give us this information, but offer tests that provide us with highly reliable data about future disease states. Of course this is not to say that there have not been dissenting voices, but even the concerns put forward by these voices seem to grant this power to genetics. For example, Jürgen Habermas opposes genetic interventions because he sees them as an affront to our bodily integrity, to our personal identities, and to our freedom. As he says,

> For as soon as adults treat the desirable genetic traits of their descendants as a product they can shape according to a design of their own liking, they are exercising a kind of control over their genetically manipulated offspring that intervenes in the somatic basis of another person's spontaneous relation-to-self and ethical freedom.[8]

This chapter will focus on some of the presumed ethical consequences of our ability to obtain genetic information about others and ourselves. One of the results of the rapid pace of genetic advances has been the increased development and use of genetic tests. Diagnostic laboratories now offer about nine hundred genetic tests, and the number of tests offered will quite likely keep growing.[9] But this new flood of genetic information presents difficult ethical decisions. The ability to screen embryos and fetuses, the capacity to test individuals at risk of transferring a genetic disease to their offspring, and the existence of presymptomatic genetic tests forces us to evaluate decisions about procreation, duties to our families, and duties to obtain genetic knowledge. This knowledge, together with the existence of reproductive technologies, has sparked a series of concerns about moral obligations to oneself and other people: from duties to know one's genetic condition, to requirements about informing family members about genetic risks, to obligations not to reproduce when one knows that offspring may have a serious genetic disease.

What, then, can biology say about the moral obligations that presumably result from the newly found ability to obtain genetic information? As we saw in previous chapters, biology has a lot to say. By paying careful attention to biological knowledge, I will show that those who argue that our ability to obtain genetic information about others and ourselves presents us with moral obligations to obtain and share that information are in error. And, again, their mistake results from a misunderstanding of the role of genetics in the development of human diseases or disorders. Such misunderstanding leads them to believe incorrectly that genetic information is sufficient to offer accurate knowledge about people's future health or lack thereof. In what follows, I will focus on three highly debated moral obligations related to the ability to acquire genetic information: the duty to know our genetic condition, the duty to inform family members about genetic risks, and the duty not to reproduce when we know that there is a high risk of transmitting a serious disease or defect.

MISUNDERSTANDING GENETIC INFORMATION

Since 1970, more than 1.4 million people worldwide have been screened voluntarily to determine whether they were carriers of the mutant gene connected to Tay-Sachs disease. Employing both enzymatic and DNA testing methods, more than 1,400 couples have been identified to be at risk of having offspring suffering from Tay-Sachs; that is, both parents carry one copy of the mutated gene. Through prenatal testing of more than 3,200 pregnancies, births of over 600 infants with this disease have been prevented.[10] Other worldwide testing or screening programs also exist for diseases such as phenylketonuria (PKU), congenital hypothyroidism, sickle-cell anemia, or beta-thalassemia. Recently, tests have started to be offered to individuals with a strong family history of breast and ovarian cancer and colon cancers. And tests have been developed that identify mutations that have been related to the development of Alzheimer's disease or cardiovascular problems.

In part due to the successful results of some of these programs, and also due to the increased knowledge of molecular genetics, a number of authors have defended the view that we have obligations to seek genetic information and use it to prevent harm. For example, some have argued that under certain circumstances we have a duty to seek information about our genetic endowment, an obligation to inform those members of our family who might also be at risk of carrying harmful genetic mutations, and a duty not to bring affected babies into this world.[11] Autonomy and beneficence have been used as grounds to defend these moral obligations. Given the availability of genetic testing, to make autonomous decisions, one ought to gain information about one's genetic endowment. Acquiring information about our genetic endowments enhances our autonomy because such information allows us to shape our lives according to our own will. According to Rosamond Rhodes,

> From a Kantian perspective, autonomy is the essence of what morality requires of me. The core content of my duty is self-determination. To say this in another way, I need to appreciate that my ethical obligation is to rule myself, that is, to be a just ruler over my own actions. As sovereign over myself I am obligated to make thoughtful and informed decisions without being swayed by irrational emotions, including the fear of knowing significant genetic facts about my self.[12]

Take, for instance, Mary. She is about to be married and is thinking about having children. If she were aware that Huntington's disease runs in her family, then knowing whether she carries the gene would allow her to make decisions about her future and the future of her family: whether to save more money; whether to have children, given that they, too, might be affected; and so on. Presumably, once she has the information she will be in a better position to decide what to do.

Similarly, some argue that our duty to prevent unnecessary harm to innocent third parties requires that we inform family members who might be at risk of being affected by genetic disorders. In the words of Walter Glannon,

> To the extent that the man had been diagnosed with Huntington's and understood the gene's degree of penetrance and the progression of the disease, he would be obligated to be tested and inform any of his children of the result. . . . The father would be obligated to be tested and to share the information because it would prevent harm in two respects. First, by having precise information about their own risks, his children might decide not to have children of their own. . . . Second, the information would better serve his children's prudential interests, enabling them to plan other aspects of their lives accordingly.[13]

Others have defended that prevention of harm imposes an obligation to avoid bringing into the world children who might suffer from a serious disability or disease. As Laura Purdy writes,

Until we can be assured that Huntington's disease does not prevent people from living a minimally satisfying life, individuals at risk for the disease have a moral duty to try not to bring affected babies into this world.[14]

Suppose that Mary, who has a sister, Tara, has just become aware that her father died of Huntington's disease. She knows now that both she and her sister might also suffer from the disease. Given the importance that such information might have for Tara's future plans, and realizing that Tara does not have any other way of finding about the possibility that she might carry the Huntington's gene, prevention of unnecessary harm would require that Mary inform her sister. Presumably, prevention of harm would also require Mary to abandon her plans to have her own children, or to use technologies that would prevent her children from being born with the Huntington's gene.

Of course, others have argued that autonomy is also grounds for a right not to be informed about one's genetic makeup. These authors contend that information about future diseases might be so distressful that it could interfere with one's ability to make rational decisions.[15] Similarly, people have argued that beneficence can ground our obligation not to inform others about their genetic risks because such information may cause psychological anguish to them.[16] These claims appear even stronger when we take into account the fact that there is no cure or possibility of prevention for many of the diseases for which genetic tests exist. Other theorists have argued that a moral obligation not to bring into the world children who might be affected by particular genetic diseases on grounds of preventing harm cannot exist because a life that contains suffering is better than no life at all. They also might argue that parents ought to unconditionally commit to any kind of child they can have.[17] Additionally, some members of the disability community have contended that to prevent the birth of children with disabilities might be harmful to people.[18] They believe that identifying fetuses that would become disabled people and choosing, because of this identification, to abort them, expresses negative judgments about disabled people and thus sends a message to living disabled people that their lives are not worth living or that they are inferior to abled people.

My concern here is not with the issue of whether autonomy and beneficence are adequate grounds to defend these moral obligations. I will assume here, for the sake of the argument, that they are. My interest is with our presumed ability to obtain reliable information about future diseases and disorders with a genetic component. Hence, the moral obligation to obtain genetic information so that one can make autonomous decisions, or the duty to inform others about their genetic makeup so that harm can be prevented, or the moral obligation not to bring children into the world who might be affected by a serious disability or disease so that suffering can be avoided all require the ability to obtain reliable information about the chances of suffering a future disease or disorder. Given the importance of obtaining reliable genetic information for a defense of these duties, it seems necessary to evaluate how reliable such knowledge is. I also want to point out that those who reject these moral obligations do so on grounds other than a concern for scientific understanding. Thus, even when they might argue that neither autonomy nor prevention of harm can justify these moral obligations, often there is no questioning of our ability to obtain highly reliable information about future disease states.

I will focus my discussion on what is normally called "predictive genetic testing" or "susceptibility testing." As we saw in chapter 6, in these kinds of tests, practitioners analyze human DNA, RNA, chromosomes, proteins, or particular metabolites to obtain information about heritable diseases. Contrary to diagnostic testing, which is used to confirm diagnosis in those who are ill, predictive genetic testing is used to identify risks in those who do not have any symptoms. In the case of prenatal testing, the detection of risks is done on those who have not even been born. These tests have as a main purpose the identification of those who are at increased risk of developing diseases or disorders later in life, such as cancer, heart problems, Alzheimer's disease, and so on.

Because we share, to different degrees, part of our genetic endowments with our relatives, information about our genetic makeup also reveals something about our family members. This is, in part, why some people have argued that the availability of these tests commits us to inform those with whom we share genetic ties. Similarly, because these tests allow us to test fetuses and embryos for future diseases, the ability to obtain such information, some have argued, imposes a duty to avoid bringing children into the world who would suffer from devastating diseases or disabilities.

Of course parents have always been faced with analyzing whether their decision to have children was a moral one in light of their particular situations, family members have needed to consider their obligations to each other, and many other kinds of knowledge about ourselves have forced us to evaluate whether we have an obligation to know. Nevertheless, the advent and continuous use of genetic technologies is providing us with knowledge about genetic diseases and disorders that parents and families did not have just a few decades ago. But the fact that we can obtain information about new diseases does not seem to be sufficient to bring about the current debate on moral obligations. After all, we have known for a long time that certain diseases or disorders "run in families." The interest in supporting moral obligations related to our ability to obtain genetic information appears to be brought about by the fact that genetic testing is thought to give us accurate knowledge about future diseases or disorders. Suppose, for example, that these tests were simply indicating a remote possibility that one might suffer in the future from a genetic disease, or that one might have a child who would suffer such a disease. If this were so, it would be hard to understand why anybody would defend the assertion that we have the kinds of moral obligations discussed here. Thus, as mentioned before, if these obligations are being defended, it must be under the assumption that genetic tests do offer us reliable information about our future health status, or that of our offspring and relatives. A misunderstanding of biology, then, grounds claims about the ethical consequences of our ability to obtain genetic knowledge.

The predictability of current genetic tests varies considerably depending on several factors. Important for the degree of predictability is, for example, whether the disease is caused by a single, highly penetrant mutation, namely, a mutation whose corresponding phenotypic trait is expressed in a high percentage of the population carrying the mutation. Also relevant is whether the disease or disorder is polygenetic or multifactorial, that is, caused by the interaction of multiple genetic and environmental factors. Obviously, the predictability of genetic tests in these two cases can be quite different. Huntington's is an example of a disorder for which genetic testing can give us highly reliable information. On the other hand, diseases such as cancer belong to the second category. Given the need for accurate knowledge, it is not surprising that those who have defended these moral obligations often have done so by using cases of genetic diseases for which genetic testing is not just available, but for which the information so obtained is highly reliable. Huntington's disease and Tay-Sachs are two of the most used cases.

Huntington's disease is due to a mutation in a gene that is transmitted as an autosomal dominant trait.[19] Each child of a parent who is carrying a mutated allele (one of a set of alternative forms of a gene) for Huntington's disease has a 50 percent risk of inheriting it. The child needs only one copy of the gene or allele from either parent to develop the disease. A person who inherits the gene involved in Huntington's disease, and who survives long enough, will sooner or later develop the disease. If the child does not inherit the defective gene, the child will not likely get the disease or pass the gene on to subsequent generations.[20] Notice that affected people normally have an affected parent, but the fact that the onset of the disease is after childbearing prevents people, unless they are tested, from realizing they have the gene before they pass it on to their children.

Tay-Sachs has a different mode of inheritance.[21] This disease is inherited in an autosomal recessive pattern, which means two copies of the gene must be altered for a person to be affected by the disorder. Although carriers of

Tay-Sachs are healthy, they have a 50 percent chance of passing on the mutated gene to their children. A child who inherits one mutated allele is a Tay-Sachs carrier like the parent. If both parents are carriers and their child inherits the mutated allele from each of them, the child will have Tay-Sachs disease. When both parents are carriers, each child has a 25 percent chance of having Tay-Sachs disease and a 50 percent chance of being a carrier.

Because Tay-Sachs disease is very rare in the general population,[22] authors using Tay-Sachs as their clinical case to support claims about particular moral obligations tend to present the case in the context of communities that have a significant number of Tay-Sachs carriers. That is the case for the Ashkenazi Jewish population (Jews of Central or Eastern European descent), in which the disease incidence is 1 in every 3,500 newborns: approximately one hundred times higher than in the general population. Among Ashkenazi Jews, 1 in every 29 individuals is heterozygous, and thus asymptomatic, for mutations causing Tay-Sachs disease. Non-Jewish French Canadians living near the St. Lawrence River and in the Cajun community of Louisiana also have a higher incidence of Tay-Sachs. The current tests detect about 95 percent of carriers of Ashkenazi Jewish background. Therefore, the predictive power of the test for Ashkenazi Jews is quite high.

It appears, then, that if genetic testing can identify the mutations involved in Huntington's disease and Tay-Sachs, then we can have highly reliable information about whether someone will suffer these diseases. This is so because a single gene that has a large effect on the phenotype controls these so-called Mendelian or monogenic disorders; that is, a single mutated gene causes them. Furthermore, these types of genetic diseases show simple patterns of inheritance within families.

The Online Mendelian Inheritance in Man (OMIM) database now records more than one thousand loci (position of a gene on a chromosome) at which there is at least one disease-causing allele; the protein affected by the mutation is also known for many genes in the OMIM database.[23] For many of these disorders we can now offer predictive and diagnostic genetic testing. It would seem, then, that given the fact that genes are being identified and analyzed for mutations, and given the fact that highly reliable tests are being provided for many of these disorders, it is reasonable to defend the position that autonomy and prevention of harm would require that we try to obtain this information about ourselves and our offspring and that we inform those members of our family who might be affected by such information.

But here is where misunderstandings about the role of genes in human diseases play their part. The genetic information about Huntington's disease or Tay-Sachs that can be obtained through genetic testing is fairly unrepresentative of the kind of information that can be obtained, and presumably will be obtainable, for the majority of diseases and disorders with a genetic component that might affect most human beings.[24] That is, most such diseases do not follow as simple a pattern of inheritance as single-gene diseases. In fact, such information is not even representative of the kind of knowledge that can be acquired by genetic testing for many of the so-called single-gene diseases. Although some traits, such as Huntington's and Tay-Sachs, are still considered to be inherited in a relatively simple monogenic manner, that is, with individual alleles segregating into families according to Mendelian laws, recent research points to a decrease in the number of disorders for which the phenotype can now be satisfactorily explained by mutations of a single allele.[25] It appears that the view of diseases as monogenic might be an oversimplification. It seems now evident that many so-called monogenic diseases have widely different phenotypes that can be accounted for by a multiplicity of mutations in a single gene and that there is a blurring of predicted relationships between genotype and phenotype in several monogenic disorders. Also, researchers now have evidence of the existence of modifier genes and nongenetic factors that contribute to the phenotypic expression of monogenic disorders.[26] For example, cystic fibrosis, one of the disorders often mentioned in the bioethics literature in discussions of some of the moral obligations discussed here, is a disease that was initially thought to be a simple monogenic disorder. This disease affects mainly the pulmonary system, but patients diagnosed with cystic fibrosis can also suffer gastrointestinal problems, defects in the pancreas, and infertility in males. Respiratory failure is the most common cause of death in cystic fibrosis patients. The median age of survival for a person with cystic fibrosis is about thirty-three years. Observations that the disease appeared to be inherited in an autosomal recessive manner led to the mapping of the gene CFTR (cystic fibrosis transmembrane conductance regulator) in 1989. After the CFTR gene was cloned, people thought that a genetic test would be sufficient to predict the clinical outcome of patients. However, we now know that, although mutations in the CFTR gene almost always cause the cystic fibrosis phenotype—that is, people with one of these mutations suffer from cystic fibrosis—modification effects by other genetic factors and probably also interaction with environmental factors prevent us from being able to predict accurately what will be the severity of the disease.[27] Different degrees of gravity of disease have been observed between individuals with identical CFTR mutations, even within the same family. Moreover, because genetic tests are directed toward finding only the mutations that are most common among the more than eight hundred mutations that have been detected for the CFTR gene, a negative diagnosis of cystic fibrosis is not possible.

Similarly, phenylketonuria, an inborn error of metabolism, which if left untreated causes mental retardation and other neurologic symptoms, is now seen as a more complex disease than it originally appeared.[28] The disease is caused by lack of the enzyme phenylalanine hydroxylase (PAH), which converts the amino acid phenylalanine to tyrosine. Often characterized as one of the classic examples of Mendelian diseases, the mapping and cloning in 1983 of the PAH gene confirmed the multiplicity of mutations of this gene. Since then, several studies have showed the existence of extensive variation of disease severity even in the presence of identical genotypes.[29] Consequently, although the inheritance of the mutant alleles does follow a Mendelian segregation pattern, the identification of the genes cannot predict the phenotype of the patient. As with cystic fibrosis, this suggests that genetic factors and the environment might be important modulating agents of the disease.

As we can see by these examples, it is difficult to defend a simple correlation between genotype and phenotype even for the so-called monogenic diseases. A given mutation may give rise to a disease with varying degrees of seriousness, and a particular mutation may be involved in several distinct diseases. Environmental influences and modifying genes may modulate the expression of a key protein. Hence, it is becoming more and more obvious that monogenic diseases are in fact more complex than they originally appeared.

Contrary to the rarity of Mendelian diseases, multifactorial ones are quite common. In fact, most of the diseases affecting human beings are of this kind. Cardiovascular diseases, cancers, autoimmune disorders, neurodegenerative diseases, and nutritional disorders — all multifactorial diseases — represent the major cause of morbidity and mortality in Western societies. However, if, as we have just seen, there is a challenge to predicting the contribution of genes to monogenic diseases, this challenge is even more pressing when we deal with multifactorial or complex diseases.

Multifactorial diseases or disorders result from mutations occurring simultaneously in several genes.[30] Unlike Mendelian diseases, the transmission of these diseases is governed by multiple factors, and familial patterns of inheritance do not follow a strictly Mendelian mode. This is due to the fact that, although each individual genetic factor involved segregates according to Mendelian laws, they do so independently of one another. Moreover, alleles contributing to these complex diseases are neither necessary nor sufficient to cause the particular disease; that is, some people might suffer the disease without having the related mutations, and some people might carry the mutations but not have the disease in question. For many of these complex diseases, more than one gene at different loci contributes to the disease, and those loci might interact with each other. Depending on their roles in the pathogenesis of diseases, these interactions might be additive, multiplicative, or have no additional effect. Modifier genes can also interact with mutations involved in the production of some diseases. The effects of interaction between an allele that might predispose toward having a particular disease and a protective allele might be especially difficult to predict with any accuracy. Similarly, as we saw before, epigenetic factors can modify the expression patterns of genes without altering the DNA sequence. The expression of most human diseases involves, as well, the relations of multiple genetic and environmental factors. Additionally, cases of incomplete penetrance and variable expressivity introduce difficulties in our ability to obtain accurate information about the risks of developing a particular disease. The phenomenon of variable expressivity explains why a mutated gene may produce phenotypes of various severities, while the phenomenon of incomplete penetrance explains the frequency with which the phenotype is seen in people who have the particular mutation.[31] The different penetrance of mutations is not entirely an intrinsic character. On the contrary, it appears to depend on several factors, such as the importance of the function of the protein encoded by the gene, the functional importance of the mutation, the interactions with other genes, the interactions with the environment, the onset of the disease, and the existence of alternative pathways that can substitute for the lost function. What is more, some of these factors can vary between individuals.[32] Things are, then, not as simple as they sometimes are made to appear.

If all the complex factors mentioned earlier were not enough to make us aware of the difficulties of obtaining highly reliable genetic information about most human diseases and disorders, there is more. Not surprisingly, despite grandiloquent reports by the press, the identification of the genes contributing to these diseases has been slow and challenging. It is true that dozens of genes have been associated in this way with diabetes, cardiovascular disease, cancer, retinitis pigmentosa, deafness, and many other diseases. Recent tabulations report about 150 genes associated with obesity in humans and in laboratory animals. However, a common experience has been that different mapping studies identify a number of apparently relevant loci, but the findings are difficult to replicate, and the initial effect estimates are biased upward.[33]

Supporters of moral obligations concerning genetic information[34] might disagree that they are misunderstanding human biology. They can certainly argue that they recognize that the strength of the obligation depends on the possibility of obtaining highly predictive information of the development of a particular disease. They might maintain that although current tests for genetic conditions might not be highly predictive of whether someone is going to develop a particular disease, this might be due to the fact that we are lacking sufficient knowledge. Consequently, they might contend, my criticisms are only pointing out a technical problem with genetic testing. Once our knowledge increases, we will be able to map and clone many of the mutations that underlie common diseases. When this is done, we will be able to offer genetic tests that reliably identify those mutations and thus that can predict future health status.

This objection is, in part, correct. Some of the problems I have pointed out have to do with insufficient knowledge and inadequate technology at this time. For example, our current tests might be unable to check for all possible mutations contributing to a particular disease. However, eventually we might be able to test for all of those mutations. If so, the reliability of the information acquired would improve significantly. Nevertheless, although the problem I am discussing here has something to do with our technological capabilities and our knowledge, it is not just a technical issue. My goal here is to call attention to the complexity of biology in general, and human biology in particular. It is this complexity that seems to have been lost in debates about the presumed moral obligations that result from our ability to obtain genetic knowledge about others and ourselves. If this is so, especially for common diseases, then uncertainty about the development of a particular disorder will arise even if we are able to identify reliably the mutations contributing to the disease.

As we have seen, complex diseases involve environmental interactions and gene-gene interactions and can require also the assessment of protective genotypes acting against the emergence of the disease phenotype. Hence, the genetic mutations underlying diseases are not the sole determinants of the disease. They might not even be the main ones. We know, for instance, that for diseases such as type II diabetes or premature atherosclerosis, environmental factors such as dietary habits, physical exercise, and obesity may all be involved before manifestation of the disease occurs.[35] In fact, type II diabetes, which affects about fifteen million Americans, is rare in the absence of obesity. And when a person suffering from type II diabetes returns to a normal body mass index, the disorder of abnormal blood sugar disappears. Consequently, for the majority of people, obesity might be as good a predictor of the disease as any genetic test. Similarly, HLA B27 is a genetic marker found in more than 90 percent of people suffering from ankylosing spondylitis, a painful, progressive, rheumatic disease. A test for HLA B27 would be, however, of quite limited use in predicting the disease because approximately 10 percent of the healthy European population also carries this genetic marker.[36]

Risk predictions on the basis of genetic data alone can give us information about only one of the many parts playing a role in the system. So, even if the tests provide highly reliable information about the genetic component, and some of the available tests certainly do, still we will not be able to do much more than what we are already able to do with other types of medical testing.[37]

Proponents of these moral duties can still dispute my claim that they misunderstand biology. They might argue that the genetic information they refer to when defending moral duties that presumably follow from our ability to obtain genetic knowledge is information that we can expect will be highly reliable. This is why, they can maintain, they tend to use cases such as Huntington's and Tay-Sachs. For these diseases, as mentioned earlier, genetic information is highly predictive of the disease phenotype. Limiting our moral obligations to obtain and share genetic information or to avoid having offspring affected by severe genetic diseases to those cases in which we can obtain reliable data about future disease would certainly solve some of the problems mentioned here. There are, however, several problems with this objection. First, even if it were the case that those who defend these moral duties do intend to limit their applicability to those diseases for which we can obtain reliable genetic information, the fact is that authors rarely point out that their defense of these obligations is very limited and directed to only a very particular and rare set of diseases, and thus to a small number of people. Although it is true that in many cases they use these very rare diseases, this seems to be the case because they are aware of the importance of reliable information in order to defend such duties. But in most cases there is no mention that the discussion is very limited in its applicability to the lives of most human beings, even to most people living in industrialized countries.

Second, as mentioned before, although it is the case that monogenic diseases are often used in these kinds of discussions, frequently authors use other diseases, such as Alzheimer's disease and breast cancer, to debate these moral obligations.[38] These diseases do not have, as we said before, the presumed clear-cut characteristics of monogenic disorders. It is true that when these clinical cases are used, they are presented in a very oversimplified way. But this only underscores the point I am making. Had the authors presented the cases in all of their complexity, they would have had difficulties maintaining that the moral obligations they propose follow.

When these cases are used, the information given about the tests and the genetic diseases is either presented in a simplistic way or is simply incorrect.[39] For example, we can be informed that testing for genetic mutations in breast cancer and Alzheimer's is available and that having such genetic mutations puts people at an increased risk of suffering the disease. In some cases we might be told that in the case of breast cancer two main mutations have been found in the BRCA1 and BRCA2 genes, which entail a significantly high risk of breast cancer (between 50 and 85 percent) and ovarian cancer (up to 60 percent) over a lifetime. Also, we are often informed that, as a dominant disorder, children of people with these mutations have a 50 percent chance of carrying the mutation. Sometimes we might also be informed that these mutations account for only about 5 to 10 percent of all breast cancers. In the case of Alzheimer's disease, authors might say that a mutation in the gene for beta-amyloid precursor protein (BAPP), when combined with the gene for apolipoprotein E (APOE), has been implicated as a cause of Alzheimer's disease. People with the BAPP mutation and one of the alleles (the e4 allele) of APOE develop Alzheimer's much earlier than those with BAPP and other alleles of APOE. People with two copies of the APOE4 have about a 90 percent chance of eventually developing the disease. But, in some cases, we are simply told that genetic testing for mutations for breast cancer and for Alzheimer's exists.

Presented in this way, we might get the impression that the predictive power of genetic tests for these conditions is quite high, and thus that the genetic information is rather deterministic. But, as mentioned earlier, this impression would be erroneous. Let us begin with the breast cancer case. BRCA1 and BRCA2 have been identified as two of the genes associated with hereditary breast and ovarian cancer.[40] This constitutes only 5 to 10 percent of all breast and ovarian cancers. More than several hundred different specific mutations have been described in the two BRCA genes thus far. Some of these mutations seem to be more prevalent in particular ethnic groups. Moreover, the gene penetrance varies with the particular mutations, and this of course impacts the ability to accurately estimate cancer risk. The reliability of BRCA screening depends on several factors, such as prior probability of the condition, sensitivity and specificity of the screening techniques, and number of mutations.[41] Recent studies have also found that nongenetic factors greatly influence cancer risk even for people with these mutations. Thus, breast cancer risk is higher in women born after 1940 than in those born before that date. In fact, the effect of birth year is much larger than that of any reported modifier gene.[42]

As for Alzheimer's disease, a neurodegenerative disease that is currently not preventable or curable, studies have documented the presence of autosomal dominant mutations on chromosomes 1, 14, and 21 in families with individuals affected by Alzheimer's, and genetic tests have been developed to test for these mutations.[43] But because these mutations are quite likely not the only mutations influencing Alzheimer's disease, the meaning of a negative test result is of limited predictive use.[44] On the other hand, a positive result must be interpreted in light of the effect of variable gene expression and incomplete penetrance on the disease development and severity. That is, having the particular mutation does not give information of when the disease will occur or how severe it will be. These facts point to limits in the predictability of tests for Alzheimer's disease. Similarly, although the presence of the APOE e4 allele is associated with a higher risk and decreased age of onset of Alzheimer's, the absolute magnitude of the risk is difficult to assess. This is so because the risk varies as a function of age, gender, ethnicity, exposure to toxins, head trauma, and other susceptibility genes.[45] Therefore, even when the test correctly reveals the presence of mutations related to Alzheimer's disease, it does not do a good job of predicting who will get the disease. A substantial proportion of APOE e4 carriers, including those surviving into their tenth decade, do not become demented, while more than one-third of persons with Alzheimer's disease do not have the e4 allele. Studies have shown that the rate of false positives and false negatives is quite high. Thus, 71 percent of individuals with one or two APOE e4 alleles never develop Alzheimer's disease, and 44 percent of people with the disease do not have the APOE e4 allele.[46]

No less important is the fact that the information available at present for common multifactorial diseases, such as cancer, cardiovascular problems, and neurodegenerative disorders, is entirely derived from very high-risk families, and thus the possible gene penetrance in other contexts, such as the general population, is unknown. Moreover, persons in whom mutations related to these diseases are found are likely to receive additional surveillance to detect possible development of the disorder. Such supervision is likely identifying cases at an earlier age. Thus, such practices might tend to increase the apparent risk of the disease.[47]

CONCLUSION

As we saw in chapter 5 with the case of human cloning, biology, when properly understood, can tell us a great deal. An analysis of the scientific mistakes present in many of our discussions about biology and its implications for our moral and public policy claims is obviously important. In the case of genetic testing and the presumed moral obligations that result from it, such an analysis can bring to our attention the fact that a correct comprehension of the workings of human biology cannot support duties to obtain information about our genetic endowment, to inform other family members, or to avoid bringing affected children into the world. These putative obligations might follow from our ability to obtain genetic information about others and ourselves only if we incorrectly presuppose that individual genes by themselves are relevant in the development of human diseases and disorders. We have seen here that such is not the case.

Discussions about genetic technologies and genetic information are often presented in ways that indicate that the predictive ability of genetic analysis is higher than what is actually defensible. This is questionable for a variety of reasons. First, as this chapter shows, to incorrectly assume that information

about our genes is highly predictive of future disease states has lead many
ethicists to defend mistakenly the notion that we have particular moral obligations to ourselves and other people—from duties to know our genetic condition, to requirements about informing family members about genetic risks,
to obligations not to reproduce when we know that our offspring may have a
genetic disease.

Second, these discussions might add support to the ideology of genetic determinism. They might help support inadequate perceptions about the in-
evitability of all diseases and disorders with a genetic component. This inadequate perception might in turn contribute to the development of public
policies that promote unnecessary use of genetic testing and screening,
preimplantation diagnoses, or genetic selection of embryos for complex disorders, and to the limitation of research efforts aimed at primary prevention.
The idea that genes are the main contributors of human diseases could also
encourage a lack of individual self-care. Already, some studies indicate that
individuals found to be at risk of heart disease through the use of a genetic
test are more likely to feel the development of the disease is inevitable and to
think that less can be done about it than do those whose assessment comes
from clinical tests.[48] Consequently, to the extent that we believe that genetic
determinism is not just false but highly problematic, bioethicists need to be
careful not to present ethical discussions in ways that might promote it.

Third, by analyzing diseases and disabilities as if they were the result exclusively of the play of our genes and as completely independent of our social life, we can contribute to the discrimination against already disadvantaged groups in our society. Moreover, we can miss the opportunity to
improve the aspects of our social, political, and legal systems that need to be
improved.

Notice that the arguments presented here are not, however, intended to defend the claim that genetic testing is never useful, that it should not be used,
or that the predictive value of current genetic tests for complex diseases will
never be improved. Certainly there are cases in which genetic testing helps
people make more informed decisions about reproduction and health care options. My point here has been to emphasize the need to take biology seriously,
to appreciate the fact that human biology, in particular, is more complex than
it appears to be in many of the discussions related to genetic information and
genetic technologies. Debates on these issues make it difficult to believe that
one of the jobs of philosophers working on bioethical issues is, as with the
mythical gadfly, to awaken people from their complacent dreams. The presumed powers of genetics have been embraced without paying much-needed
careful attention to some of the customary beliefs and assumptions about the
role of science in society; about the presumed value neutrality of scientific

knowledge; and, of course, about the assumed deterministic role of genes in human disease and behavior.

Nevertheless, the defense of these moral obligations is dubious on grounds other than scientific blunders. Thus, even if the arguments that I have presented here are incorrect, and even if we could hope to obtain accurately predictive genetic information about future health states, still these moral obligations might be indefensible. This is so because the defense of these moral obligations does not just tend to embrace a particular conception of the role of genes in human disease, but it also accepts that we can adequately discuss these moral obligations by looking at the biology and ignoring the social context in which these new moral obligations would confront human beings. However, if we reject this understanding of moral obligations and put such duties into context, we can see that their defense is not so straightforward. This will be the focus of the next chapter. As in the case of human cloning, although biology can tell us a lot, it cannot tell us everything we need to consider when dealing with moral and public policy consequences of biological knowledge.

NOTES

1. See, for example, F. Collins and V. McKusick, "Implications of the Human Genome Project for Medical Science," *JAMA* 285, no. 5 (2001): 540–44; J. Watson, *A Passion for DNA: Genes, Genome, and Society* (New York: CSHL Press, 2000); W. Gilbert, "A Vision of the Grail," in *The Code of Codes: Scientific and Social Issues in the Human Genome Project*, ed. D. J. Kevles and L. Hood, 83–97 (Cambridge, MA: Harvard University Press, 1992); T. Lee, *The Human Genome Project: Cracking the Genetic Code of Life* (New York: Plenum, 1991).

2. F. Collins and V. McKusick, "Human Genome Project."

3. See, for example, L. Walters and J. G. Palmer, *The Ethics of Human Gene Therapy* (New York: Oxford University Press, 1997).

4. See L. Kass, "The Wisdom of Repugnance," in *Flesh of My Flesh*, ed. G. Pence, 13–37 (Lanham, MD: Rowman & Littlefield, 1998); U.S. National Bioethics Advisory Commission, *Cloning Human Beings: Report and Recommendations of the National Bioethics Advisory Commission* (Rockville, MD: The Commission, 1997); A. Verhey, "Cloning: Revisiting an Old Debate," *Kennedy Institute of Ethics Journal* 4 (1994): 227–34; D. Callahan, "Perspective on Cloning: A Threat to Individual Uniqueness," *Los Angles Times*, November 12, 1993, B7. See also chapters 3–5 for references on this topic.

5. See, for example, E. Juengst, "Can Enhancement Be Distinguished from Prevention in Genetic Medicine?" *Journal of Medicine and Philosophy* 22, no. 2 (1997): 125–42; J. Glover, *What Sort of People Should There Be?* (New York: Penguin, 1984).

6. See, for example, C. Diver and J. Cohen, "Genophobia: What Is Wrong with Genetic Discrimination?" *University of Pennsylvania Law Review* 149, no. 5 (2001):

1439–82; P. Roche and G. Annas, "Protecting Genetic Privacy," *Nature Reviews: Genetics* 2, no. 5 (2001): 392–96; K. Hudson et al., "Genetic Discrimination and Health Insurance: An Urgent Need for Reform," *Science* 270, no. 5235 (1995): 391–93.

    7. See, for example, R. Rhodes, "Genetic Links, Family Ties, and Social Bonds: Rights and Responsibilities in the Face of Genetic Knowledge," *Journal of Medicine and Philosophy* 23, no. 1 (1998): 10–30.

    8. J. Habermas, *The Future of Human Nature* (Cambridge, UK: Polity, 2003), 13.

    9. F. S. Collins, *A Brief Primer on Genetic Testing* (National Human Genome Research Institute, 2003). Available at http://www.genome.gov/10506784 (accessed 16 Feb. 2005).

    10. M. Kaback, "Population-Based Genetic Screening for Reproductive Counseling: The Tay-Sachs Disease Model," *European Journal of Pediatrics* 159, suppl. 3 (2000): S192–S195.

    11. W. Glannon, *Genes and Future People* (Boulder, CO: Westview Press, 2001), chap. 2; A. Buchanan, D. Brock, N. Daniels, and D. Wikler, *From Chance to Choice: Genetics and Justice* (Cambridge, UK: Cambridge University Press, 2000), chap. 6; H. Clarkeburn, "Parental Duties and Untreatable Genetic Conditions," *Journal of Medical Ethics* 26, no. 5 (2000): 400–405; A. Sommerville and V. English, "Genetic Privacy: Orthodoxy or Oxymoron?" *Journal of Medical Ethics* 25, no. 2 (1999): 144–50; R. Rhodes, "Genetic Links"; D. Davis, "Genetic Dilemmas and the Child's Right to an Open Future," *The Hastings Center Report* 27, no. 2 (1997): 7–15; L. Purdy, "Genetics and Reproductive Risk: Can Having Children Be Immoral?" in *Reproducing Persons: Issues in Feminist Bioethics* (Ithaca, NY: Cornell University Press, 1996), 39–49; B. Steinbock and R. McClamrock, "When Is Birth Unfair to the Child?" *The Hastings Center Report* 24, no. 6 (1994): 15–21.

    12. R. Rhodes, "Genetic Links," 18.

    13. W. Glannon, *Genes and Future People*, 44.

    14. L. Purdy, "Can Having Children Be Immoral?" 49.

    15. See T. Takala and M. Häyry, "Genetic Ignorance, Moral Obligations and Social Duties," *Journal of Medicine and Philosophy* 25, no. 1 (2000): 107–13; T. Takala, "The Right to Genetic Ignorance Confirmed," *Bioethics* 13, nos. 3–4 (1999): 288–93; A. Huibers and A. van 't Spijker, "The Autonomy Paradox: Predictive Genetic Testing and Autonomy; Three Essential Problems," *Patient Education and Counseling* 35, no. 1 (1998): 53–62.

    16. See, for example, T. Takala and M. Häyry, "Genetic Ignorance"; T. Takala, "Right to Genetic Ignorance."

    17. See S. Vehmas, "Parental Responsibility and the Morality of Selective Abortion," *Ethical Theory and Moral Practice* 5, no. 4 (2002): 463–84; S. Vehmas, "Just Ignore It? Parents and Genetic Information," *Theoretical Medicine* 22, no. 5 (2001): 473–84.

    18. See, for example, L. Carlson, "The Morality of Prenatal Testing and Selective Abortion: Clarifying the Expressivist Objection," in *Mutating Concepts, Evolving Disciplines: Genetics, Medicine and Society*, ed. L. S. Parker and R. A. Ankeny, 191–213 (Dordrecht, the Netherlands: Kluwer Academic Publishers, 2002); E. Parens and A. Asch, eds., *Prenatal Testing and Disability Rights* (Washington, DC: Georgetown University Press, 2000).

19. See, for example, C. A. Ross and R. L. Margolis, "Huntington's Disease," *Clinical Neuroscience Research* 1, nos. 1–2 (2001): 142–52; L. Ho et al., "The Molecular Biology of Huntington's Disease," *Psychological Medicine* 31, no. 1 (2001): 3–14; C. L. Wellington et al., "Toward Understanding the Molecular Pathology of Huntington's Disease," *Brain Pathology* 7, no. 3 (1997): 979–1002; A. Wexler, *Mapping Fate* (Berkeley, CA: University of California Press, 1996); Huntington's Disease Collaborative Research Group, "A Novel Gene Containing a Trinucleotide Repeat That Is Expanded and Unstable on Huntington's Disease Chromosomes," *Cell* 72, no. 6 (1993): 971–83.

20. Some cases of Huntington's disease in individuals with no family history of the disease have been found. See C. Holzmann et al., "Functional Characterization of the Human Huntington's Disease Gene Promoter," *Molecular Brain Research* 92, nos. 1–2 (2001): 85–97; A. Durr et al., "Diagnosis of 'Sporadic' Huntington's Disease," *Journal of the Neurological Sciences* 129, no. 1 (1995): 51–55.

21. See, for example, N. Risch, "Molecular Epidemiology of Tay-Sachs Disease," *Advances in Genetics* 44 (2001): 233–52; R. Rozenberg and V. Pereira, "The Frequency of Tay-Sachs Disease Causing Mutations in the Brazilian Jewish Population Justifies a Carrier Screening Program," *Sao Paulo Medical Journal* 119, no. 4 (2001): 146–49; L. Peleg et al., "Mutations of the Hexosaminidase A Gene in Ashkenazi and Non-Ashkenazi Jews," *Biochemical Medicine and Metabolic Biology* 52, no. 1 (1994): 22–26; S. Yamanaka et al., "Structure and Expression of the Mouse Beta-Hexosaminidase Genes, *Hexa* and *Hexb*," *Genomics* 21, no. 3 (1994): 588–96.

22. See, for example, V. R. Sutton, "Tay-Sachs Disease Screening and Counseling Families at Risk for Metabolic Disease," *Obstetrics and Gynecology Clinics of North America* 29, no. 2 (2002): 287–96; R. E. Zinberg et al., "Prenatal Genetic Screening in the Ashkenazi Jewish Population," *Clinical Perinatology* 28, no. 2 (2001): 367–82; G. M. Petersen et al., "The Tay-Sachs Disease Gene in North American Jewish Populations: Geographic Variations and Origins," *American Journal of Human Genetics* 35, no. 6 (1983): 1258–69; R. Rozenberg and V. Pereira, "Frequency of Tay-Sachs."

23. C. R. Scriver, "Why Mutation Analysis Does Not Always Predict Clinical Consequences: Explanations in the Era of Genomics," *Journal of Pediatrics* 140, no. 5 (2002): 502–6.

24. See, for example, M. Cummings, *Human Heredity: Principles and Issues*, 5th ed. (Pacific Grove, CA: Brooks/Cole, 2000), chap. 5; A. Wilkie, "Genetic Prediction: What Are the Limits?" *Studies in the History and Philosophy of Biological and Biomedical Sciences* 32, no. 4 (2001): 619–33; N. Risch, "Searching for Genetic Determinants in the New Millennium," *Nature* 405 (2002): 847–56.

25. J. L. Baldano and N. Katsanis, "Beyond Mendel: An Evolving View of Human Genetic Disease Transmission," *Nature Reviews: Genetics* 3, no. 10 (2002): 779–89.

26. C. R. Scriver and P. J. Waters, "Monogenic Traits Are Not Simple: Lessons from Phenylketonuria," *Trends in Genetics* 15, no. 7 (1999): 267–72.

27. R. A. Ankeny, "Reduction Reconceptualized: Cystic Fibrosis as a Paradigm Case for Molecular Medicine," in *Mutating Concepts, Evolving Disciplines: Genetics, Medicine and Society*, ed. L. S. Parker and R. A. Ankeny, 127–41 (Dordrecht, the Netherlands: Kluwer Academic Publishers, 2002); K. M. Dipple and E. McCabe, "Modifier

Genes Convert 'Simple' Mendelian Disorders to Complex Traits," *Molecular Genetics and Metabolism* 71, nos. 1–2 (2000): 43–50; M. Stuhrmann et al., "Mutation Screening for Prenatal and Presymptomatic Diagnosis: Cystic Fibrosis and Haemochromatosis," *European Journal of Pediatrics* 159, supp. 3 (2000): S186–S191; J. E. Mickle and G. R. Cutting, "Genotype-Phenotype Relationships in Cystic Fibrosis," *The Medical Clinics of North America* 84, no. 3 (2000): 597–607.

28. For a review of the history of PKU screening in the United States, see D. Paul, "The History of Newborn Phenylketonuria Screening in the U.S.," app. 5 in *Promoting Safe and Effective Genetic Testing in the United States*, ed. N. A. Holtzman and M. S. Watson, 137–59 (Washington, DC: NIH-DOE Working Group on the Ethical, Legal, and Social Implications of Human Genome Research, 1997).

29. J. L. Baldano and N. Katsanis, "Beyond Mendel"; C. R. Scriver and P. J. Waters, "Monogenic Traits Are Not Simple."

30. M. Cummings, *Human Heredity: Principles and Issues*, chap. 5; A. Wilkie, "Genetic Prediction"; N. Risch, "Genetic Determinants."

31. C. Dennis, "Epigenetics and Disease: Altered States," *Nature* 421, no. 6924 (2003): 686–89; A. Wilkie, "Genetic Prediction"; N. Risch, "Genetic Determinants."

32. P. Vineis et al., "Misconceptions about the Use of Genetic Tests in Populations," *Lancet* 357, no. 9257 (2001): 709–12.

33. D. J. Galton and G. A. Ferns, "Genetic Markers to Predict Polygenic Disease: A New Problem for Social Genetics," *QJM* 92, no. 4 (1999): 223–32.

34. See note 11 for references.

35. See, for example, L. Quinn, "Behavior and Biology," *The Journal of Cardiovascular Nursing* 18, no. 1 (2003): 62–68; R. R. Wing et al., "Behavioral Science Research in Diabetes," *Diabetes Care* 24, no. 1 (2001): 117–23; D. J. Galton and G. A. Ferns, "Genetic Markers"; and M. Rewers and R. F. Hamman, "Risk Factors for Non-Insulin-Dependent Diabetes," in *Diabetes in America* (Bethesda, MD: NIH, 1995), 179–220.

36. D. J. Galton and G. A. Ferns, "Genetic Markers."

37. K. Weiss and A. Buchanan, "Evolution by Phenotype: A Biomedical Perspective," *Perspectives in Biology & Medicine* 46, no. 2 (2003): 159–82.

38. See, for example, W. Glannon, *Genes and Future People*; L. d'Agincourt-Canning, "Experiences of Genetic Risk: Disclosure and the Gendering of Responsibility," *Bioethics* 15, no. 3 (2001): 231–47; R. Wachbroit, "The Question Not Asked: The Challenge of Pleiotropic Genetic Tests," *Kennedy Institute of Ethics Journal* 8, no. 2 (1998): 131–44.

39. See, for example, W. Glannon, *Genes and Future People*; A. Buchanan, D. Brock, N. Daniels, and D. Wikler, *From Chance to Choice*, chap. 6; R. Wachbroit, "The Question Not Asked."

40. See, for example, P. L. Welcsh and M. C. King, "BRCA1 and BRCA2 and the Genetics of Breast and Ovarian Cancer," *Human Molecular Genetics* 10, no. 7 (2001): 705–13.

41. See, for example, M. J. Coughlin, "Ethical Implications of Genetic Testing for Breast Cancer Susceptibility," *Critical Reviews in Oncology/Hematology*, 40, no. 2 (2001): 149–57; S. S. Khoury and K. K. Steinberg, "BRCA1 and BRCA2 Gene Muta-

tions and Risk of Breast Cancer: Public Health Perspectives," *American Journal of Preventive Medicine* 16, no. 2 (1999): 91–98; N. Weitzel, "The Current Social, Political, and Medical Role of Genetic Testing in Familial Breast and Ovarian Carcinomas," *Current Opinion in Obstetrics and Gynecology* 11, no. 1 (1999): 65–70; J. Garber, "A 40-Year-Old Woman with a Strong Family History of Breast Cancer," *JAMA* 282, no. 20 (1999): 1953–60; J. L. Hopper et al., "Population-Based Estimate of the Average Age-Specific Cumulative Risk of Breast Cancer for a Defined Set of Protein-Truncating Mutations in BRCA1 and BRCA2: Australian Breast Cancer Family Study," *Cancer Epidemiology, Biomarkers & Prevention* 8, no. 9 (1999): 741–47; D. Ford et al., "Genetic Heterogeneity and Penetrance Analysis of the BRCA1 and BRCA2 Genes in Breast Cancer Families," *American Journal of Human Genetics* 62, no. 3 (1998): 676–89; H. Welch and W. Burke, "Uncertainties in Genetic Testing for Chronic Disease," *JAMA* 280, no. 17 (1998): 1525–27; F. Collins, "BRCA1—Lots of Mutations, Lots of Dilemmas," *New England Journal of Medicine* 334 (1996): 186–88.

42. E. Levy-Lahad and S. Plon, "A Risky Business: Assessing Breast Cancer Risk," *Science* 302, no. 5645 (2003): 574–75.

43. See, for example, S. L. Tobin et al., "The Genetics of Alzheimer Disease and the Application of Molecular Tests," *Genetic Testing* 3, no. 1 (1999): 37–45.

44. L. Bertram and R. E. Tanzi, "Alzheimer's Disease: One Disorder, Too Many Genes?" *Human Molecular Genetics* 13, spec. no. 1 (2004): R135–R141; M. I. Kamboh, "Molecular Genetics of Late-Onset Alzheimer's Disease," *Annals of Human Genetics* 68, pt. 4 (2004): 381–404.

45. See, for example, R. C. Green, "Risk Assessment for Alzheimer's Disease with Genetic Susceptibility Testing: Has the Moment Arrived?" *Alzheimer's Care Quarterly* 3, no. 3 (2002): 208–14; L. A. Farrer et al., "Effects of Age, Sex, and Ethnicity on the Association between Apolipoprotein E Genotype and Alzheimer Disease: A Meta-Analysis," *JAMA* 278, no, 16 (1997): 1349–56; J. Kahn, "Ethical Issues in Genetic Testing for Alzheimer's Disease," *Geriatrics* 52, no. 9 (1997): 30–32; S. G. Post et al., "The Clinical Introduction of Genetic Testing for Alzheimer's Disease," *JAMA* 277, no. 10 (1997): 832–36; T. D. Bird et al., "Wide Range in Age of Onset for Chromosome 1–Related Familial Alzheimer Disease," *Annals of Neurology* 40, no. 6 (1996): 932–36; A. D. Roses, "Apolipoprotein E Affects the Rate of Alzheimer's Disease Expression: Beta-Amyloid Burden Is a Secondary Consequence Dependent on APOE Genotype and Duration of Disease," *Journal of Neuropathology and Experimental Neurology* 53, no. 5 (1994): 429–37.

46. A. Wilkie, "Genetic Prediction."

47. P. Vineis et al., "Misconceptions about the Use of Genetic Tests in Populations"; H. Welch and W. Burke, "Uncertainties in Genetic Testing for Chronic Disease."

48. V. Senior, T. Marteau, and J. Weinman, "Impact of Genetic Testing on Causal Models of Heart Disease and Arthritis: An Analogue Study," *Psychology & Health* 14, no. 6 (2000): 1077–88; V. Senior, T. Marteau, and T. Peters, "Will Genetic Testing for Predisposition for Disease Result in Fatalism? A Qualitative Study of Parents' Responses to Neonatal Screening for Familial Hypercholesterolaemia," *Social Science and Medicine* 48, no. 12 (1999): 1857–60.

Chapter Eight

Moral Obligations, Genetic Information, and Social Context

In chapter 7, I argued for a need to recognize and embrace the complexity of human biology when evaluating the consequences of biological knowledge for our moral life. That chapter informed us of what biology can tell us. And, what it tells us is that a correct understanding of human biology can help us assess whether we can support obligations to obtain genetic information about ourselves, to inform our family members, and to avoid bringing particular children into this world.

This chapter, on the other hand, calls attention to what biology cannot tell us. Here, I will show that even if we grant that genetic tests do give us highly predictive information about our future health status, still we might have difficulties espousing these moral obligations. This is so because the defense of such obligations presupposes the legitimacy of discussing moral duties in a decontextualized and abstract fashion. This account, as we will see, posits universal obligations independently of the social contexts in which such obligations would be binding for human beings. It discounts the relevance of particulars such as gender, economic class, or ethnicity so as to offer principles that presumably would be universally applicable. Such an understanding of moral obligations, I will show, is questionable for several reasons. First, it is unhelpful to the real humans beings to which those moral obligations presumably apply. Second, and more importantly, by neglecting the social and political context in which people make moral decisions, we run the risk of reinforcing or furthering injustices in an already unjust world. If we face social and political institutions that hinder particular groups or individuals and we then propose moral obligations that because of the unjust institutional context would place greater burdens on, or further disadvantage, those groups or individuals, we can contribute to reinforcing such injustices. Third, a defense of particular moral obligations might demand that we strive to transform those social structures that make the exercise of such moral duties unjustly burdensome to particular individuals or groups. If we ignore the social context in which people make moral decisions, we will do little to ensure the change of those structures that perpetuate injustices against some people.

GENETIC INFORMATION AND MORAL OBLIGATIONS: WHAT BIOLOGY CANNOT TELL US

Reliance on highly abstract, universal principles as the appropriate source of moral guidance is one of the trademarks of the Western philosophical tradition. In spite of the many differences between deontologists and utilitarians, these approaches share the view that moral problems are to be solved by the application of these abstract principles to cases. But some of the many problems with this understanding of moral reasoning are that particulars often get ignored, the relevance of the networks of human relationships neglected, the importance of intimacy obscured, and the broad social and political arrangements in which moral decision making takes place overlooked.

During the last few decades, a number of criticisms of mainstream ethical theory have appeared or resurfaced.[1] Of these criticisms, the ones coming from a feminist perspective have been especially concerned with the need for attending to the social and political context.[2] They have called attention to the inadequacy of the impartial and universalizing character of traditional ethical theories when it comes to dealing with issues of intimate relationships; situated people; or the influence of gender, race, ethnicity, and class inequalities in ethical decision making. In spite of these criticisms, the defense of abstract moral rules and the decontextualization of moral problems have not withered.

Although I will be dealing here with only a few issues that are relevant when dealing with presumed moral obligations related to obtaining genetic information, this does not mean that such problems are the only ones in need of consideration. Many other aspects of our social structures, political institutions, public policies, and so on also need to be taken into account in order to understand the effects that the defense of such duties might have on real, situated human beings. For example, concerns about the possibility of genetic discrimination are certainly relevant for any discussion of putative moral obligations resulting from our ability to obtain genetic information about others or ourselves. Whether people might have difficulties in obtaining insurance coverage because of their genetic status or in accessing particular kinds of jobs is significant if we are going to argue that people have a moral duty to acquire information about their genetic makeup and an obligation to inform affected relatives. Because the issue of genetic discrimination has received considerable attention both from academics and from public policy makers, I will not discuss it here.[3] Similarly, no less relevant for a discussion of moral obligations that presumably result from our newly found ability to obtain genetic knowledge is the effect that such obligations might have on disabled people. Although we need more awareness of the problems that disabled people face in general, and of the effects that the new genetic technologies might have on this group, important work in this area has been forthcoming.[4]

My reasons for focusing on the particular problems I discuss here have to do in part with the inadequate attention they have received in bioethical discussions related to genetic technologies. Defending moral obligations to obtain genetic information about ourselves, to inform family members about genetic risks, and to not reproduce when we know there is a high risk of transmitting a serious disease or defect requires the use of medical technologies and health care services. Hence, it seems that attention to who does and does not have access to genetic testing and genetic counseling, to how the offering of these services might affect health care systems, and to what other medical priorities our societies might have is particularly relevant. Given the increases in health care costs and the lack of insurance coverage for a great part of the human population, to neglect these issues seems questionable. Also, considerations of how much the public and health professionals know about genetic issues and of whether people can understand test results in ways that might enhance their decision-making abilities have been neglected, but they seem significant for a discussion of moral duties that presumably result from new genetic technologies. Similarly, addressing the need for attending to gender when evaluating moral obligations related to genetic information is justified because of the following reasons. First, recent analysis of the role of evaluative assumptions in scientific knowledge has shown that biology seems particularly situated to impact women's well-being. For example, during the early nineteenth century, biological theories of reproduction encouraged public policies that prevented women from attending school.[5] Also, research on sex hormones seems to underlie the medical emphasis on hormone replacement therapy during menopause in spite of evidence of an increased risk of cancer.[6] Second, except in limited cases, evaluations of the possible effects on women of new scientific and technological advances and of moral obligations are scarce.[7] This is in spite of the fact that the few cases in which gender issues are addressed clearly show that, in many instances, advances in genetic science and technology seem to have a different and more troublesome impact on women. This is so because of women's role in reproduction, their greater responsibility in child rearing, and their underrepresentation in the health care professions.[8]

Access to Genetic Technologies

Common to the presumed moral obligations under discussion is that they all require the use of particular technologies in order to be fulfilled. To obtain information about our own genetic makeup so that we can make autonomous decisions and inform family members who might be affected by such information, we need genetic testing. To fulfill a moral obligation not to bring children into the world who might be affected by a serious genetic disease, several available technologies are also needed. Thus, as we saw in chapter 6, people might choose to do carrier testing to determine whether they are at risk of transmitting a harmful genetic mutation. Women might also decide to have prenatal testing or opt for preimplantation diagnosis with in vitro fertilization.

The first difficulty we find when we discuss these moral obligations in the current social context has to do with access to genetic testing. The costs of a genetic test vary from less than $100 to more than $2,000 depending on the complexity of the test, the number of individuals tested to obtain a result that is meaningful, and the method of specimen handling.[9] For example, depending on the methodology used, costs of tests for breast cancer mutations BRCA1 and BRCA2 range from $350 to $1,290, tests for hereditary non-polyposis colorectal cancer range from $250 to $3,000, tests for familial adenomatous polyposis range from $235 to $1,000, and tests for cystic fibrosis cost between $200 and $250. Moreover, genetic counseling is usually recommended and sometimes required before and after genetic testing is performed. Genetic counseling for breast cancer, for example, prior to testing costs on average $213, whereas counseling, testing, and disclosure of results costs around $2,000.[10]

In the case of trying to fulfill a moral duty not to bring children with serious genetic conditions into the world, a decontextualized discussion of this moral obligation ignores still other important factors. For example, if a pregnancy is already in course, we have issues about whether women lacking insurance, or those who do not possess adequate coverage, have access to prenatal testing. Further, if the test results indicate that the fetus carries a deleterious genetic mutation, we must also consider the availability of abortion services. To ignore this issue is quite troublesome given that not only do many women lack the ability to pay for such services, and the federal government does not allow use of federal funding for abortions, but also because many jurisdictions limit access to second- and third-term abortions. Relevant also to the possibility of abortion because the fetus might have a disability is whether some women might feel pressured to have abortions. Given the negative images presented about a life with disabilities and the lack of social support, it is not unreasonable to believe that, in this context, many women might feel forced to choose aborting a fetus diagnosed with a possible disability.[11]

Still more factors must be considered if a pregnancy is desired and abortion is not seen as an option. In this case, in order to fulfill a presumed moral obligation not to bring affected children into the world, people can decide to use preimplantation genetic diagnosis (PGD) and IVF. The use of these technologies has particular implications for women that will be discussed in a later section. But it also has economic consequences because these technologies are quite expensive, and most insurers do not cover them. The average cost of an IVF cycle ranges from $10,000 to $43,000, excluding the costs of previous treatments and the postnatal expenses.[12] PGD can add $2,500 to $4,000.[13]

When paying attention to the social contexts in which people make moral decisions, we can realize that a requirement to use medical technologies in order to fulfill one's moral obligations is troublesome. For example, in the United States, health expenditures reached $1.5 trillion in 2002. This constitutes nearly 15 percent of the GDP.[14] In spite of this, in 2003, over 45 million people in the country, or 15.6 percent of the total population, had no health care coverage.[15] The poor and the uninsured often cannot get access to minimally adequate health care. Thus, although Medicaid insured 14 million people in poverty in 2002, over 10 million other people in poverty had no health insurance.[16] Additionally, the number and percentage of people covered by employment-based health insurance is decreasing, while private insurance coverage is becoming more and more expensive.

The inadequacy of health care access affects minority populations considerably. Hence, Hispanic persons and non-Hispanic black persons are more likely to lack health insurance than are non-Hispanic white persons.[17] The fact that the U.S. health insurance system is mainly based on employer contributions also affects minority populations significantly. This is so because minorities have higher levels of unemployment and more often work part time, in service jobs, or in temporary jobs that prevent them from having adequate health benefits. For example, in early 2005, the unemployment rate for Hispanics was 2.7 percent higher, and for African Americans over 6 percent higher, than for white people.[18] Additionally, poverty rates for minority households are much higher than in the population as a whole. Thus, over 22 percent of Hispanic families and over 24 percent of blacks were below the poverty line in 2003. The percentage for non-Hispanic whites was 8.2 percent.[19] Statistics from the U.S. Census Bureau for 2003 note that the median household income for blacks was $29,645, and for Hispanics it was $32,997, while for non-Hispanic whites it was $47,777.[20]

Even for those who do have insurance, access to genetic testing is not guaranteed. At present, state laws with regard to access to, and coverage of, genetic services are limited to newborn screening and childhood genetic diseases. No state requires health insurance coverage of genetic testing for adult onset disorders, such as breast cancer, which, as mentioned earlier, may cost more than a thousand dollars.[21] Group health insurance plans often do not cover screening tests in the absence of symptoms and thus exclude coverage of genetic testing for many diseases. Where coverage exists for genetic testing, reimbursement is rarely provided for the necessary education and counseling that should accompany such tests, especially with respect to prenatal diagnosis. In addition, if genetic predisposition to a disease is detected, the patient may seek expensive periodic screening or prophylactic surgery.[22] In countries like the United States, then, with a large private-health sector, receipt of genetic testing and related services depends mainly on the individual's ability to pay.

In countries with universal health care coverage, problems of access to genetic testing are less prominent. Such access would certainly go a long way in helping people to meet their putative moral duties to obtain information about their genetic endowment and to assure that they will not bring into the world children who might suffer from serious genetic conditions. Nevertheless, genetic testing, genetic counseling, new specialists, research facilities, educational programs, and the like all need public and private investments, and thus not many countries currently offer such services. In a world of limited resources, where a multiplicity of diseases, not just genetic ones, affect human beings, and where access to diagnosis and treatment is limited for many people, countries must set health care priorities. Research, development, and use of genetic testing and related services have opportunity costs. Consequently, money used for these techniques cannot be used for other health-related needs. Investment in development of, and access to, these technologies needs therefore to be evaluated in relation to many other, and in some cases more pressing, health needs. This is especially the case when we deal with universal access to genetic testing for conditions for which no treatment is available. Similarly, as services for genetic testing expand beyond rare familial disorders to broader populations, cost impacts for public health systems will tend to grow. Although, of course, universal access to predictive genetic testing could be cost effective, analyses of this matter are still lacking. A decontextualized discussion of moral obligations that presumably result from our ability to obtain genetic information about ourselves and our offspring would inevitably neglect the fact that such analyses are necessary.

Understanding Genetic Information

Dealing with moral obligations related to genetic information in a decontextualized fashion does not just overlook economic considerations. It also neglects issues related to people's ability to understand the science underlying genetic testing and the results of genetic tests. Similarly, it overlooks issues related to whether there are resources available to help people with such tasks.

National and international studies reveal that the public is not well educated about genetic concepts or about available genetic technologies. Studies of schoolchildren and young people, those who have attended genetic counseling where an explanation of the relevant genetics is generally given, and the public at large show that, despite widespread efforts to popularize Mendelian genetics throughout the century, public understanding of it is very limited.[23] Recent research has called attention to the fact that lack of knowledge about risk factors is pervasive across all different ages, races, socioeconomic groups, and particularly among the disadvantaged.[24] Studies about public perception of cancer risks indicate that people misconceive the relationship between age and cancer risk, for example, and tend to believe that cancer risks decrease rather than increase with age.[25] Research on breast cancer also indicates that although people might be aware of the importance of family history in disease susceptibility, still they are unclear about how or from whom the cancer could be inherited. A common misperception is that breast cancer risk could only be inherited from the mother's side of the family.[26] Studies of community knowledge about cystic fibrosis carrier screening and prenatal diagnosis found very poor knowledge of the disorder and even more limited awareness of its inheritance pattern.[27] Some research has suggested that Mendelian explanations of inheritance are poorly accepted and understood because they conflict in a number of ways with a widespread lay understanding of inheritance that is derived from the social relationships of kinship. It might be that people do not find the Mendelian concept easy because it does not fit with what they already believe.[28]

Moreover, as we saw in the previous chapter, genetic tests give us information about the presumed risks of suffering a particular disease. It is well-known, however, that both experts and laypeople have difficulties calculating and understanding probabilities of risk and risk-related information, especially when that information is presented to them in a quantitative fashion.[29] For example, people tend to believe that events are more probable when they can recall an incident of its occurrence. They also deem that in normal circumstances they are less likely than their peers to suffer harm.[30] Additionally, individuals' understanding of numeric data appears to be informed by particular cognitive biases.[31] For instance, in a study on muscular dystrophy, researchers found that although the subjects had been given risk values of their carrier status and reproductive risk expressed in percentage terms, mothers and daughters of families with Duchenne muscular dystrophy frequently translated those values into ordinal or descriptive statements that were appropriate for reproductive decision making. Thus, rather than using percentages, they described themselves as having a high risk, a low risk, as being a carrier, as not being able to have boys, and so on. Such translations turned the inherently probabilistic notion of risk into definitive categories.[32] Studies also show that recipients of genetic information related to reproductive concerns reduce the information to a dichotomous interpretation, that is, to either it will or will not happen.[33] Additionally, people are unfamiliar with uncertainty in risk assessment, and risk attitudes affect perceptions of uncertainty.[34] Psychological elements, including individuals' preconceptions, misconceptions, and previous experiences, will thus influence how they construe risk information. Characteristics of people's general worldview also influence how risk information is interpreted. For example, individuals described as pessimists may overestimate the significance of a particular risk value.[35]

Some studies suggest that people have difficulties understanding the results of genetic tests or whether the tests predict risk or diagnose disease.[36] For instance, some of the tests used to detect cystic fibrosis carrier status can identify only 85 percent of mutations, and thus a negative result indicates that there is a risk of 1 in 160 of carrying a mutation. In spite of this, 17 to 29 percent of people with a negative test think, shortly after taking the test, that they are definitively not carriers.[37] Another recent study on understanding results of tests for cystic fibrosis mutations reports that 23 percent of carriers falsely believed they were only very likely to be carriers, while nearly a third of test-negative individuals falsely believed they were definitely not carriers.[38]

Help from medical geneticists and genetic counselors could go a long way toward trying to solve the problems of lack of knowledge and difficulty understanding test results and risk information. However, there are several problems, in addition to the problems of access discussed above, with this possible solution. Although available data indicate that the numbers of individuals graduating from human genetics training programs are increasing, it is not clear that this increase is occurring at the rate necessary to ensure adequate and appropriate levels of support for genetic services in the future.[39] Due to the relative scarcity of professionals specially trained to deal with genetic aspects of health, disease, treatment, and prevention, there is a widespread belief that primary care providers, whether physicians, nurses, or other health care professionals, as well as nongeneticist specialist physicians, will be the ones providing the necessary genetic services to their patients.[40]

For these services to be effective, primary care providers should have time to provide thorough counseling or to engage in lengthy discussions with patients. They should also have knowledge of important genetic disorders and patterns of inheritance, genetic testing procedures and availability, and possible therapies and treatments. Moreover, practitioners should also have skills in understanding and communicating risk information. Unfortunately, studies show that primary care physicians think they lack knowledge of genetics, the nature of inherited disorders, and screening techniques and availability.[41] They also believe that there have been inadequate educational opportunities to learn about genetics and genetic counseling. Many of them miss opportunities for genetic diagnosis and offer few referrals to genetic counseling services. In some cases, primary care physicians are unaware of the availability of prenatal diagnosis for a variety of genetic disorders, and of carrier testing. They also lack the skills to collect a genetic family history. As a result, most primary care practitioners are not confident in offering advice. They also report difficulties with the integration of genetic services into their current practices.[42]

Invisible Women

Decontextualized defenses of moral obligations related to our ability to obtain genetic information also present these moral obligations in gender-neutral terms. This is seriously problematic because of the presence of social, economic, and political institutions related to health and reproduction that systematically disadvantage women. Obligations to obtain and disclose genetic information might affect women differentially for several reasons. First, traditionally women have being responsible for family health. Women bear a disproportionate burden for the health care of children and partners. They tend to them when they are ill, negotiate professional health care for their families, and communicate health care information on their behalf.[43] Hence, although caregiving is often framed in gender-neutral terms, in practice, women are the ones expected to bear the main responsibility for family care. And this is the case both in the private and the public realm.[44] Wives and mothers are seen as having a duty to care for their husbands and children, as well as for other family members. Health care reform in many countries has also increased demands on women to take on caregiving responsibilities that were previously managed by professionals.[45] Given that women are the primary caregivers for children and other family members, an obligation to seek and disclose genetic information about themselves would impose disproportional burdens on them. This obligation would be seen as a natural extension of their caregiving duties. Thus, they would be the ones in charge of delivering to their relatives what many might see as bad news. This might produce conflicts with women's perceived duty of giving care because the disclosure of information could result in their relatives being anxious about the possibility of developing particular diseases. Women might then see themselves as causing others harm. Likewise, women would be faced with making decisions about which family members might need to be informed, how to deliver the information, or when the disclosure should occur.[46]

Another reason why obligations to obtain and disclose genetic information might affect women differentially is related to the fact that conceptions of responsibility are influenced by social meanings and practices.[47] In the context of our social practices, we have evidence that suggests that women tend to see the self not as an atomized, autonomous agent, but as constructed in relation to others, as an interdependent self.[48] They see their lives as interconnected with the lives of others and define themselves in terms of their social relationships with others and their obligations to them. In this context, to ignore the effects on women that the moral obligations to inform themselves or others might have is seriously questionable. Women seem to be prepared to undergo potentially risky medical interventions to fulfill their perceived obligation of caring for others.[49] Some studies on women undergoing genetic testing or attending genetic counseling indicate that women see the search for genetic information as a way to help their relatives. They tend to see their role in generating genetic information for their relatives as the right thing to do. In many cases they cite the need to preserve others' autonomy, often at the expense of their own, as the justification to obtain information about their own genetic endowments.[50] This raises questions about whether women's search for genetic information might be constrained by their perceived need to care for and help their relatives, as well as by societies' perceptions that women are more adequate than men to care for others. Moreover, women who are identified as at risk of suffering some genetic condition also assume responsibility for managing such risks. Consequently, some women justify their willingness to adopt potentially harmful risk management, such as prophylactic mastectomies, by appealing to their responsibilities to fulfill their role as mothers or to prevent relatives from seeing them suffer or having to care for them.[51] Some research also shows that women, more so than men, may view any potential risk as a negative appraisal of themselves.[52]

Given that society in general, and women in particular, tend to see women as responsible for maintaining their own health and the health of their relatives, to propose moral obligations to generate genetic information and to share it with family members might increase the burdens that women already bear. In many cases, women's decisions not to use genetic testing might be considered irresponsible behavior.[53] Hence, by ignoring the social and political context in which these moral obligations take place and thus how they might affect women differentially, we neglect the fact that justice requires that the burdens as well as the benefits of obtaining and disclosing genetic information be distributed fairly.

Because women are often seen as more responsible for reproductive decision making, and because genetic testing is often presented as a way for women to exercise reproductive autonomy,[54] defending moral duties not to reproduce when we know that there is a high risk of transmitting a serious disease or defect is likewise troubling. As we mentioned earlier, in order to fulfill the duty not to reproduce when we know that there is a high risk of transmitting a serious disease or defect, people might act in several ways. First, they might choose, or be pressured,[55] to do carrier testing to determine whether they are at risk of transmitting a harmful genetic mutation. Second, if the pregnancy has already occurred, women might decide to have prenatal testing. Third, if a pregnancy is desired and abortion is not seen as an option, preimplantation diagnosis might be possible. In all of these cases, a moral obligation not to reproduce when we know there is a high risk of transmitting a serious disease or defect disproportionately burdens women. Thus, to present such a moral duty in a decontextualized, gender-neutral manner might foster injustices against them.

Although at first glance the decision to undergo carrier testing for reproductive reasons might seem to affect women and men equally, in actuality this is far from being the case. Research indicates that women tend to accept offers of free carrier testing for particular genetic conditions (e.g., cystic fibrosis) more than men in population-based screening programs.[56] Women also tend to be tested initially when there might be a risk of transmitting a genetic mutation to their children. Only if the woman tests positive for a recessive disorder is the male partner involved in the testing.[57] Some have suggested that this might be due to the fact that men see carrier testing as related to reproductive choices and thus more of a woman's responsibility.[58] It might also be related to the sense of responsibility for and to others that women tend to have. Other researchers have indicated that the fact that more women than men accept carrier testing might be related to the fact that women tend to use primary health care services more than men.[59] In any case, lack of attention to the social context in which we are proposing the fulfillment of moral obligations would neglect that more women than men undergo carrier testing.

Once pregnancy occurs, the fulfillment of a moral duty not to reproduce when we know that there is a high risk of transmitting a serious disease or defect requires that we make use of prenatal testing. Currently there are three basic approaches to identify fetal defects within the women's womb: visualization (e.g., ultrasonography), analysis of fetal tissues (e.g., amniocentesis, chorionic villus sampling), and laboratory studies (e.g., DNA analysis, cytogenics). All these approaches allow doctors to determine whether fetuses have certain chromosomal abnormalities, such as Down's syndrome, and sex-chromosome disorders. These techniques also enable practitioners to check for genetic mutations such as Tay-Sachs disease and Huntington's disease.

Although these tests present different degrees of intrusiveness, obviously any information about a fetus's genetic condition is mediated through the woman's body. Presenting moral obligations not to reproduce under particular circumstances in a decontextualized fashion neglects the fact that only women can undergo amniocentesis, ultrasound, or any other of these tests. Furthermore, prenatal testing presents the possibility of choosing whether or not to terminate the pregnancy based on the result of the test, and women are the ones who have abortions. As mentioned before, to this we must add the difficulty that many women find, in some jurisdictions, of obtaining late termination after detection of a genetic abnormality following amniocentesis. Thus, compared to men, women have to accept more physical invasion and also more responsibility for their fetuses and children.

The fact that women are the only ones who can undergo these kinds of procedures does not make the moral obligation in question problematic. There are, however, systemic and structural factors that disadvantage pregnant women and mothers. For example, problems related to lack of maternity leave, childcare services, access to health care, or employment opportunities present women with serious difficulties that men generally do not face. These hindrances do not result from the facts of pregnancy or motherhood, but from social and political choices that necessitate that when women take on the responsibilities associated with pregnancy and motherhood, they do so without adequate social and institutional support. Hence a defense of a decontextualized moral duty not to bring particular children into the world might further injustices against women by increasing the physical and emotional burdens of pregnancy and motherhood.

If a pregnancy is desired, but a couple knows, or suspects, that they might transmit a harmful genetic mutation and they don't see abortion as an option, they can fulfill their moral duty not to bring an affected child into the world by using preimplantation genetic diagnoses. The advent of IVF has made possible the creation of embryos in the laboratory. These embryos can now be tested through preimplantation genetic diagnoses for a number of genetic diseases. [60] Of course, IVF is a set of procedures that can only be undergone by women. Furthermore, IVF involves physical and emotional costs. [61] Women must learn an extraordinary amount of information to adequately prepare for IVF treatment. They must learn how to mix and administer injectable medications, interrupt their daily routines for serial blood tests and ultrasound examinations, and undergo a surgical procedure to retrieve oocytes. Moreover, IVF might pose serious risks to women's health. According to empirical evidence, risks to women undergoing IVF treatment range from simple nausea to death. For example, the hormones that doctors use to stimulate the ovaries are associated with numerous side effects. Some studies assert that ovulation induction may be a risk factor for certain types of hormone-dependent cancers.[62] A substantial body of experimental, clinical, and epidemiological evidence indicates that hormones play a major role in the development of several human cancers. The ability of hormones to stimulate cell division in certain organs, such as the breast, the endometrium, and the ovary, may lead (following repeated cell divisions) to the accumulation of random genetic errors that ultimately produce cancer. Hormone-related cancers account for more than 30 percent of all newly diagnosed female cancer in the United States.[63] Hence, any technique (like IVF) that relies on massive doses of hormones may be quite dangerous.

The ovarian hyperstimulation syndrome (OHSS) is another possible iatrogenic (caused by medical treatment) consequence of ovulation induction.[64] Women with the severe form of OHSS may suffer renal impairment, liver dysfunction, thromboembolic phenomena, shock, and even death. The incidence of moderate and severe OHSS in IVF treatment ranges from 3 to 4 percent. This seems quite a high risk, taking into account that IVF is an elective procedure. This syndrome is extremely rare following natural conception. Additionally, the procedures that doctors normally use to obtain women's eggs (i.e., laparoscopy and ultrasound-guided oocyte retrieval) also pose risks to women.[65] Although there are no accurate statistical data about hazards associated with these two procedures, risks related to these technologies include postoperative infections, punctures of an internal organ, hemorrhages, ovarian trauma, and intrapelvic adhesions. Implantation of embryos or gametes into women's bodies also may be hazardous for them.[66] Some of the possible risks are perforation of organs and ectopic pregnancies. Studies show that 2 to 5 percent of all IVF pregnancies implant outside the uterus. The hazard in the general population, however, is approximately 1 percent. Ectopic gestations can also be life threatening for the woman.

The fulfillment of a moral obligation not to bring affected offspring into the world when people might wish to have children and they might be at risk of transmitting a serious genetic disease, again, overburdens women. As mentioned before, the injustices of this disproportionate burden are not due to the fact that only women can undergo these procedures. Inattentiveness to social context, however, can blind us to how, because of existing unjust social structures and public policies, a biological fact such as getting pregnant can disadvantage women.

SOME OBJECTIONS AND RESPONSES

My criticism of a decontextualized defense of moral obligations related to seeking, disclosing, and using genetic information might raise several objections.

First, as a general objection, critics might argue that the moral duties to obtain genetic information about oneself, to disclose it to other affected parties, and to not bring affected children into the world are prima facie obligations. As such, these obligations need to be considered in the context of the full scope of people's responsibilities. This being the case, it seems unnecessary to employ lengthy discussions about the social context in which these moral obligations are put into practice. Because these obligations are prima facie, individuals ought to make judgments about their particular situations. They must take into account the social context in which such obligations take place, whether the fulfillment of such duties might impose excessive economic or emotional burdens on them, whether they have competing moral duties, what political implications their actions might have, and so on. To the extent that their evaluations are correct, individuals who refuse to seek genetic information, disclose it to relatives, or use it for reproductive purposes are fulfilling what they take to be their duties. Thus, for example, only those individuals who can pay for genetic services, either with their own money or through their public or private insurances, are morally required to use them in order to fulfill their presumed moral duties.

This objection is, however, questionable. Take for example the case of access to genetic services. It seems strange, to say the least, to defend a moral obligation that, because of present economic conditions, we know cannot be met by a significant part of the population. It would seem that we are taking as the norm for our moral duties a particular economic standard that is available to only some people. But if we are going to do so, we should provide arguments for why, in a context where most people do not meet such an economic standard, we should regard these presumed moral duties as such. Absent these arguments, when people are regularly excused for doing what we otherwise would consider wrong (to not obtain genetic information about themselves or about their offspring), they are then dismissed as individuals who are incapable of behaving morally. This might create conditions for guilt for those individuals unable to perform their duties, even when they could be reasonably excused from doing so. Most people, aware of their moral obligations, and realizing that they are unable to fulfill them, would feel blameworthy because they are not doing what would be thought morally right. Add to this the fact that many individuals could suffer moral condemnation because in many cases it is difficult to know whether someone has good reasons to disregard their obligations.

Appealing to a prima facie nature of moral duties related to genetic information thus begs the question. It seems to assume that a decontextualized evaluation of moral obligations related to genetic information is an adequate one, as long as we recognize that these are only prima facie obligations. My point here has been to argue that such is not the case. First, to conceive of these moral obligations in this way ignores the fact that such duties could only arise in certain, very specific contexts: those of societies with very advanced medical technologies. Outside of this particular social context, these obligations would have no meaning whatsoever. Second, if we think that obligations related to genetic information do exist, to try to present them in decontextualized ways is inadequate. When we try to analyze these moral obligations by simply pointing out that other things need to be considered, we are implicitly and uncritically sanctioning the status quo. Considering these obligations in context would draw our attention to the structural problems that must be transformed if we are to fulfill such duties. For example, as pointed out earlier, the presumed duties to obtain and share genetic information, and the putative obligation to not bring children into the world who might suffer from serious genetic conditions, disproportionately burdens women. One could say that if these are really obligations, the fact that women are over-burdened seems insufficient to reject the existence of the obligations. And this might be the case. However, when we ignore the particular context in which women are burdened with these obligations, we fail to identify and then transform those factors that are responsible for the disproportionate effects on women. Thus, we contribute to further injustices against them. It seems unclear then how an appeal to the prima facie character of these obligations would solve the problem pointed out here, that is, the possibility of increasing injustices against those who already bear the greatest burden of fulfilling these moral duties or against those who are excluded from access to genetic services in order to carry them out.

A particular criticism to my arguments that women, in trying to fulfill these obligations, would bear a disproportionate burden is that social roles often impose differential obligations on people. Hence, we might expect women to have obligations that emanate from their roles as mothers, sisters, or caregivers. This observation seems correct. However, nothing I have said here denies this point. People do, and ought to, have particular obligations that stem from their roles as parents, teachers, government officials, doctors, midwives, gardeners, and the like. Meeting these duties can, of course, place different burdens on people, and this need not be unjust or unfair. Nevertheless, social choices can either assist people in meeting their obligations or can obstruct their doing so. My goal here is to point out that in an institutional context in which women already face a variety of obstacles to carrying out the duties resulting from their social roles, to propose additional obligations, while ignoring the effects of the social context, can exacerbate existing injustices.

CONCLUSION

As we have seen in this chapter, the evaluation of the moral and public policy consequences that presumably result from biological knowledge requires that we pay attention, not just to the scientific information, but also
to the present social context. Unfortunately, in many cases, such attention is
lacking.

Discussions about the alleged moral duties that follow from our ability to
obtain genetic information about others and ourselves often ignore that access
to these technologies is limited, that laypeople and professionals might not
have adequate knowledge about genetics and genetic testing technologies,
and that already disadvantaged groups might be unfairly burdened by these
obligations. Although the universalizing character of moral duties might give
us some insights into moral behavior, a strict adherence to an impartial, abstract, unsituated perspective appears not just unhelpful, but dangerous.
Where we believe that certain moral duties are binding, paying attention to
the social context in which such duties are implemented will allow us to encourage transformation of those social structures that make the exercise of
such moral duties difficult for many human beings. If we ignore the social
and political context in which people make moral decisions, we will do little
to ensure the improvement of those structures that currently perpetuate injustices against some people. Significantly, although as we saw in chapter 7, biology can tell us a great deal about moral and public policy issues, it cannot
tell us everything we need to know in order to properly evaluate moral claims
related to biomedical science and technologies.

NOTES

1. See, for example, M. Nussbaum, *Love's Knowledge* (Oxford, UK: Oxford Uni-
versity Press, 1990); A. Jonsen and S. Toulmin, *The Abuse of Casuistry* (Berkeley,
CA: University of California Press, 1988); A. MacIntyre, *After Virtue* (Notre Dame,
IN: University of Notre Dame Press, 1981); S. Toulmin, "The Tyranny of Principles,"
Hastings Center Report 11 (1981): 31–39; P. Foot, *Virtues and Vices* (Oxford, UK:
Blackwell, 1978).

2. See, for example, A. L. Carse, "Impartial Principle and Moral Context: Securing a Place for the Particular in Ethical Theory," *Journal of Medicine and Philosophy* 23, no. 2 (1998): 153–69; S. Wolf, ed., *Feminism and Bioethics* (New York: Oxford University Press, 1996); A. Baier, *Moral Prejudices* (Cambridge, MA: Harvard
University Press, 1994); S. Sherwin, *No Longer Patient* (Philadelphia: Temple University Press, 1992); V. Held, *Feminist Morality* (Chicago: University of Chicago
Press, 1993).

3. See, for example, D. Hellman, "What Makes Genetic Discrimination Exceptional?" *American Journal of Law and Medicine* 29, no. 1 (2003): 77–116; A. Silvers and M. A. Stein, "Human Rights and Genetic Discrimination: Protecting Genomics' Promise for Public Health," *Journal of Law, Medicine and Ethics* 31, no. 3 (2003): 377–89; U.S. Congress, Senate, Committee on Health, Education, Labor, and Pensions, *Protecting Against Genetic Discrimination: The Limits of Existing Laws* (Washington, DC: U.S. Government Printing Office, 2002); U.S. Congress, Committee on Energy and Commerce, Subcommittee on Commerce, Trade, and Consumer Protection, *The Potential for Discrimination in Health Insurance Based on Predictive Genetic Tests* (Washington, DC: U.S. Government Printing Office, 2002); M. F. Otlowski, S. D. Taylor, and K. K. Barlow-Stewart, "Major Study Commencing into Genetic Discrimination in Australia," *Journal of Law and Medicine* 10, no. 1 (2002): 41–48; M. A. Rothstein and M. R. Anderlik, "What Is Genetic Discrimination, and When and How Can It Be Prevented?" *Genetics in Medicine* 3, no. 5 (2001): 354–58; T. Lemmens, "Selective Justice, Genetic Discrimination, and Insurance: Should We Single Out Genes in Our Laws?" *McGill Law Journal* 45, no. 2 (2000): 347–12; C. A. Tauer, "Genetic Testing and Discrimination: How Can We Protect Job and Insurance Policy Applicants from Negative Test Consequence?" *Health Progress* 82, no. 2 (2001): 48–53, 71; K. L. Hudson, "Genetic Discrimination and Health Insurance: An Urgent Need for Reform," *Science* 270, no. 5235 (1995): 391–93; P. S. Miller, "Genetic Discrimination in the Workplace," *Journal of Law, Medicine and Ethics* 26, no. 3 (1998): 189–97.

4. See, for example, P. Herissone-Kelly, "Bioethics in the United Kingdom: Genetic Screening, Disability Rights, and the Erosion of Trust," *Cambridge Quarterly of Healthcare Ethics* 12, no. 3 (2003): 235–41; A. Brooks, "Women's Voices: Prenatal Diagnosis and Care for the Disabled," *Health Care Analysis* 9 (2001): 133–50; T. Koch, "Disability and Difference: Balancing Social and Physical Constructions," *Journal of Medical Ethics* 27, no. 6 (2001): 370–76; J. L. Nelson, "Prenatal Diagnosis, Personal Identity, and Disability," *Kennedy Institute of Ethics Journal* 10, no. 3 (2000): 213–28; S. M. Reindal, "Disability, Gene Therapy and Eugenics — A Challenge to John Harris," *Journal of Medical Ethics* 26, no. 2 (2000): 89–94; A. Buchanan, D. Brock, N. Daniels, and D. Wikler, *From Chance to Choice: Genetics and Justice* (Cambridge, UK: Cambridge University Press, 2000), chap. 7; E. Parens and A. Asch, *Prenatal Testing and Disability Rights* (Washington, DC: Georgetown University Press, 2000); E. Parens and A. Asch, "The Disability Rights Critique of Prenatal Genetic Testing: Reflections and Recommendations," *Hastings Center Report* 29, no. 5 (1999): S1–S22; A. Silvers, D. Wasserman, and M. Mahowald, *Disability, Difference, Discrimination* (Lanham, MD: Rowman & Littlefield, 1998); J. Fitzgerald, "Geneticizing Disability: The Human Genome Project and the Commodification of Self," *Issues in Law & Medicine* 14, no. 2 (1998): 147–63; A. Buchanan, "Choosing Who Will Be Disabled: Genetic Intervention and the Morality of Inclusion," *Social Philosophy and Policy* 13, no. 1 (1996): 18–46.

5. N. Tuana, *The Less Noble Sex: Scientific, Religious, and Philosophical Conceptions of Woman's Nature* (Bloomington, IN: Indiana University Press, 1993); C. Meigs, *Women and Their Diseases* (Philadelphia: Lea & Blanchard, 1848).

6. A. Fausto-Sterling, *Sexing the Body: Gender Politics and the Construction of Sexuality* (New York: Basic Books, 2000); I. de Melo-Martín, *Making Babies: Biomedical Technologies, Reproductive Ethics, and Public Policy* (Dordrecht, the Netherlands: Kluwer, 1998).

7. For some of the exceptions, see, for example, L. d'Agincourt-Canning, "Experiences of Genetic Risk: Disclosure and the Gendering of Responsibility," *Bioethics* 15, no. 3 (2001): 231–47; M. Mahowald, *Genes, Women, Equality* (New York: Oxford University Press, 2000); R. Hubbard, "Genetics and Women's Health," *Journal of the American Medical Women's Association* 52, no. 1 (1997): 2–3; A. Asch and G. Geller, "Feminism, Bioethics, and Genetics," in *Feminism and Bioethics: Beyond Reproduction*, ed. S. Wolf (New York: Oxford University Press, 1996); M. Mahowald, "A Feminist Standpoint for Genetics," *Journal of Clinical Ethics* 7, no. 4 (1996): 333–40; L. Purdy, "What Can Progress in Reproductive Technology Mean for Women?" *Journal of Medicine and Philosophy* 21, no. 5 (1996): 499–514; L. S. Parker, "Breast Cancer Genetic Screening and Critical Bioethics' Gaze," *Journal of Medicine and Philosophy* 20, no. 3 (1995): 313–37; A. Lippman, "Worrying—and Worrying about—the Geneticization of Reproduction and Health," in *Misconceptions: The Social Construction of Choice and the New Reproductive Technologies*, vol. 1, ed. G. Basen, M. Eichler, and A. Lippman, 39–65 (Ottawa, Canada: Voyageur Press, 1993); A. Lippman, "Prenatal Genetic Testing and Screening: Constructing Needs and Reinforcing Inequities," *American Journal of Law and Medicine*, 17, nos. 1–2 (1991): 15–50.

8. I. de Melo-Martín and C. Hanks, "Genetic Technologies and Women: The Importance of Context," *Bulletin of Science, Technology & Society*, 21, no. 5 (2001): 354–60; M. Mahowald, *Genes, Women, Equality*.

9. U.S. National Library of Medicine, National Institutes of Health, *Genetics Home Reference* (Bethesda, MD: 2004). Available at http://ghr.nlm.nih.gov/info=genetic_testing/show/cost_results;jsessionid=4F0A6C9A1FEC9D2B52043E37E479 2F48 (accessed 21 Feb. 2005).

10. W. F. Lawrence et al., "Cost of Genetic Counseling and Testing for BRCA1 and BRCA2 Breast Cancer Susceptibility Mutations," *Cancer Epidemiology Biomarkers & Prevention* 10, no. 5 (2001): 475–81.

11. See L. Carlson, "The Morality of Prenatal Testing and Selective Abortion: Clarifying the Expressivist Objection," in *Mutating Concepts, Evolving Disciplines: Genetics, Medicine and Society*, ed. L. S. Parker and R. A. Ankeny, 191–213 (Dordrecht, the Netherlands: Kluwer Academic Publishers, 2002); E. Parens and A. Asch, eds., *Prenatal Testing and Disability Rights*; L. B. Andrews and M. Hibbert, "Courts and Wrongful Birth: Can Disability Itself Be Viewed as a Legal Wrong?" in *Americans with Disabilities: Exploring Implications of the Law for Individuals and Institutions*, ed. L. Pickering Francis and A. Silver (New York: Routledge, 2000): 318–330.

12. J. Botkin, "Ethical Issues and Practical Problems in Preimplantation Genetic Diagnosis," *Journal of Law, Medicine & Ethics* 26, no. 1 (1998): 18.

13. Genetics and Public Policy Center, *Preimplantation Genetic Diagnosis* (Genetics and Public Policy Center, 2004). Available online at http://www.dnapolicy.org/downloads/pdfs/policy_pgd.pdf (accessed 31 Jan. 2005).

14. National Center for Health Statistics, *Health, United States, 2004* (Hyattsville, MD: National Center for Health Statistics, 2004), 14. Available at http://www.cdc .gov/nchs/data/hus/hus04.pdf (accessed 14 Jan. 2005).

15. C. DeNavas-Walt, B. D. Proctor, and R. J. Mills, U.S. Census Bureau, Current Population Reports, P60-226, *Income, Poverty, and Health Insurance Coverage in the United States: 2003* (Washington, DC: U.S. Government Printing Office, 2004), 14. Also available at http://www.census.gov/prod/2004pubs/p60-226.pdf (accessed 21 Jan. 2005).

16. U.S. Department of Commerce, Economics and Statistics Administration, U.S. Census Bureau, *Health Insurance Coverage in the United States: 2002* (Washington, DC: U.S. Census Bureau, 2003), 1–4. Also available at http://www.census.gov/prod/ 2003pubs/p60-223.pdf (accessed 5 Feb. 2005).

17. C. DeNavas-Walt, B. D. Proctor, and R. J. Mills, *Income*, 15.

18. U.S. Department of Labor, Bureau of Labor Statistics, News, Bureau of Labor Statistics, *The Employment Situation: January 2005* (Washington, DC: BLS, 2005), 2. Available at http://www.bls.gov/news.release/pdf/empsit.pdf (accessed 10 Mar. 2005).

19. C. DeNavas-Walt, B. D. Proctor, and R. J. Mills, *Income*, 10.

20. C. DeNavas-Walt, B. D. Proctor, and R. J. Mills, *Income*, 7.

21. A. Johnson, *Genetics and Health Insurance* (Denver, CO: NCSL, 2002). Available at http://www.ncsl.org/programs/health/genetics/Geneticshealthins.pdf (accessed 14 Feb. 2005).

22. M. A. Rothstein and S. Hoffman, "Genetic Testing, Genetic Medicine, and Managed Care," *Wake Forest Law Review* 34, no. 3 (1999): 849–88.

23. M. P. Richards, "Lay Understanding of Mendelian Genetics," *Endeavour* 22, no. 3 (1998): 93–94.

24. R. A. Breslow et al., "Americans' Knowledge of Cancer Risk and Survival," *Preventive Medicine* 26, no. 2 (1997): 170–77.

25. K. Honda and A. I. Neugut, "Associations between Perceived Cancer Risk and Established Risk Factors in a National Community Sample," *Cancer Detection and Prevention* 28, no. 1 (2004): 1–7; E. A. Grunfeld et al., "Women's Knowledge and Beliefs Regarding Breast Cancer," *British Journal of Cancer* 86 (2002): 1373–78; N. C. Dolan, A. M. Lee, and M. M. McDermott, "Age-Related Differences in Breast Carcinoma Knowledge, Beliefs, and Perceived Risk among Women Visiting an Academic General Medicine Practice," *Cancer* 80, no. 3 (1997): 413–20.

26. N. Vuckovic et al., "Consumer Knowledge and Opinions of Genetic Testing for Breast Cancer Risk," *American Journal of Obstetrics and Gynecology* 189, no. 4 (2003): S48–S53.

27. M. Decruyenaere et al., "Cystic Fibrosis: Community Knowledge and Attitudes Towards Carrier Screening and Prenatal Diagnosis," *Clinical Genetics* 41 (1992): 189–96.

28. M. P. Richards, "Lay Understanding of Mendelian Genetics."

29. See, for example, D. Kahneman, P. Slovic, and A. Tversky, eds., *Judgment Under Uncertainty: Heuristics and Biases* (Cambridge, UK: Cambridge University Press, 1974).

30. N. D. Weinstein and W. M. Klein, "Resistance of Personal Risk Perceptions to Debiasing Interventions," *Health Psychology* 14, no. 2 (1995): 132–40.

31. J. L. Bottorff et al., "Communicating Cancer Risk Information: The Challenges of Uncertainty," *Patient Education and Counseling* 33, no. 1 (1998): 67–81.

32. E. Parsons and P. Atkinson, "Lay Constructions of Genetic Risk," *Sociology of Health and Illness* 14, no. 4 (1992): 437–55. See also N. Hallowell, H. Statham, and F. Murton, "Women's Understanding of Their Risk of Developing Breast/Ovarian Cancer Before and After Genetic Counseling," *Journal of Genetic Counseling* 7, no. 4 (1998): 345–64.

33. A. Lippman-Hand and F. C. Fraser, "Genetic Counseling: Provision and Reception of Information," *American Journal of Medical Genetics* 3, no. 2 (1979): 113–27.

34. B. B. Johnson and P. Slovic, "Presenting Uncertainty in Health Risk Assessment: Initial Studies of Its Effects on Risk Perception and Trust," *Risk Analysis* 15, no. 4 (1995): 485–94.

35. J. L. Bottorff et al., "Communicating Cancer Risk Information."

36. N. Vuckovic et al., "Consumer Knowledge and Opinions."

37. C. Bankhead et al., "New Developments in Genetics–Knowledge, Attitudes and Information Needs of Practice Nurses," *Family Practice* 18, no. 5 (2001): 475–86.

38. C. Gordon et al., "Population Screening for Cystic Fibrosis: Knowledge and Emotional Consequences 18 Months Later," *American Journal of Medical Genetics* 120, no. 2 (2003): 199–208.

39. A. L. Kinmonth, J. Reinhard, and S. Pauker, "The New Genetics: Implications For Clinical Service in Britain and the United States," *British Medical Journal* 316, no. 7133 (1998): 767–70.

40. W. Burke and J. Emery, "Genetic Education for Primary-Care Providers," *Nature Reviews: Genetics* 3, no. 7 (2002): 561–66; M. Mahowald et al., *Genetics in the Clinic: Clinical, Ethical, and Social Implications for Primary Care* (Philadelphia: Mosby, 2001).

41. See, for example, S. Metcalfe et al., "Needs Assessment Study of Genetics Education for General Practitioners in Australia," *Genetics in Medicine* 4, no. 2 (2002): 71–77; J. Emery and S. Hayflick, "The Challenge of Integrating Genetic Medicine into Primary Care," *British Medical Journal* 322, no. 7293 (2001): 1027–30; M. D. Fetters, D. J. Doukas, and K. L. D. Phan, "Family Physicians' Perspectives on Genetics and the Human Genome Project," *Clinical Genetics* 56 (1999): 28–34; J. Emery et al., "A Systematic Review of the Literature Exploring the Role of Primary Care in Genetic Services," *Family Practice* 16, no. 4 (1999): 426–45.

42. R. Robins and S. Metcalfe, "Integrating Genetics as Practices of Primary Care," *Social Science and Medicine* 59, no. 2 (2004): 223–33.

43. See M. Navaie-Waliser, A. Spriggs, and P. H. Feldman, "Informal Caregiving: Differential Experiences by Gender," *Medical Care* 40, no. 12 (2002): 1249–59; K. Donelan, M. Falik, and C. M. DesRoches, "Caregiving: Challenges and Implications for Women's Health," *Women's Health Issues* 11, no. 3 (2001): 185–200; J. Wuest, "Repatterning Care: Women's Proactive Management of Family Caregiving

Demands," *Health Care for Women International* 21, no. 5 (2000): 393–411; R. L. Hoffmann and A. M. Mitchell, "Caregiver Burden: Historical Development," *Nursing Forum* 33, no. 4 (1998): 5–11.

44. J. Parks, *No Place Like Home?* (Bloomington, IN: Indiana University Press, 2003); E. F. Kittay, *Love's Labor* (New York: Routledge, 1999).

45. J. Wuest, "Repatterning Care."

46. N. Hallowell, "Doing the Right Thing: Genetic Risk and Responsibility," *Sociology of Health and Illness* 21, no. 5 (1999): 597–621.

47. L. d'Agincourt-Canning, "Experiences of Genetic Risk"; M. Walker, *Moral Understandings: A Feminist Study in Ethics* (New York: Routledge, 1998); V. Held, *Feminist Morality.*

48. See, for example, C. Gilligan, *In a Different Voice: Psychological Theory and Women's Development* (Cambridge, MA: Harvard University Press, 1982). See also A. Baier, *Moral Prejudices*; V. Held, *Feminist Morality.*

49. N. Hallowell, "Doing the Right Thing."

50. See N. Hallowell et al., "Balancing Autonomy and Responsibility: The Ethics of Generating and Disclosing Genetic Information," *Journal of Medical Ethics* 29, no. 2 (2003): 74–79; G. Goelen et al., "Moral Concerns of Different Types of Patients In Clinical BRCA1/2 Gene Mutation Testing," *Journal of Clinical Oncology* 17, no. 5 (1999): 1595–1600; J. Mason, "Gender, Care and Sensibility in Family and Kin Relationships," in *Sex, Sensibility, and the Gendered Body*, ed. J. Holland and L. Adkins, 15–36 (Basingstoke, UK: Macmillan, 1996).

51. N. Hallowell, "Doing the Right Thing."

52. J. E. Newman et al., "Gender Differences in Psychosocial Reactions to Cystic Fibrosis Carrier Testing," *American Journal of Medical Genetics* 113, no. 2 (2002): 155; G. Evers-Kiebooms et al., "A Stigmatizing Effect of the Carrier Status for Cystic Fibrosis?" *Clinical Genetics* 46, no. 5 (1994): 336–43.

53. For a discussion rejecting charges of moral irresponsibility for parents who fail to use genetic testing technologies when they know their possible children are at risk for a serious genetic disorder, see J. Andre, L. M. Fleck, and T. Tomlinson, "On Being Genetically 'Irresponsible,'" *Kennedy Institute of Ethics Journal* 10, no. 2 (2000): 129–46.

54. See, for example, T. M. Marteau et al., "Long-Term Cognitive and Emotional Impact of Genetic Testing for Carriers of Cystic Fibrosis: The Effects of Test Result and Gender," *Health Psychology* 16, no. 1 (1997): 51–62; M. Stacey, "The New Genetics: A Feminist View," in *The Troubled Helix: Social and Psychological Implications of the New Human Genetics*, ed. T. M. Marteau and M. P. M. Richards, 331–49 (Cambridge, UK: Cambridge University Press, 1996); J. C. Callahan, ed., *Reproduction, Ethics, and the Law* (Bloomington, IN: Indiana University Press, 1995); C. Overall, *Ethics and Human Reproduction* (Boston: Allen & Unwin, 1987); B. K. Rothman, *The Tentative Pregnancy* (New York: Viking, 1986); M. O'Brien, *Politics of Reproduction* (London: Routledge Kegan and Paul, 1981).

55. See, for example, L. B. Andrews and M. Hibbert, "Courts and Wrongful Birth: Can Disability Itself Be Viewed as a Legal Wrong?" in *Americans with Disabilities:*

Exploring Implications of the Law for Individuals and Institutions, ed. L. Pickering Francis and A. Silver, 318–30 (New York: Routledge, 2000).

56. H. Bekker et al., "Uptake of Cystic Fibrosis Testing in Primary Care: Supply Push or Demand Pull?" *British Medical Journal* 306 (1993): 1584–86; E. K. Watson et al., "Psychological and Social Consequences of Community Carrier Testing Screening for Cystic Fibrosis," *Lancet* 340 (1992): 217–20.

57. M. Mahowald, *Genes, Women, Equality.*

58. H. Bekker et al., "Uptake of Cystic Fibrosis Testing."

59. E. K. Watson et al., "Consequences of Community Carrier Testing."

60. D. Wells and D. A. Delhanty, "Preimplantation Genetic Diagnosis: Applications for Molecular Medicine," *Trends in Molecular Medicine* 7, no. 1 (2001): 23–30.

61. I. de Melo-Martín, *Making Babies*, chap. 4.

62. A. Venn, "Risk of Cancer after Use of Fertility Drugs with In-Vitro Fertilization," *Lancet* 354, no. 9190 (1999): 1586–90; R. E. Bristow and B. Y. Karlan, "The Risk of Ovarian Cancer after Treatment for Infertility," *Current Opinion in Obstetric and Gynecology* 8, no. 1 (1996): 32–37; A. Shushan et al., "Human Menopausal Gonadotropin and the Risk of Epithelial Ovarian Cancer," *Fertility and Sterility* 65, no. 1 (1996): 13–18.

63. See, for example, J. V. Lacey Jr. et al., "Menopausal Hormone Replacement Therapy and Risk of Ovarian Cancer," *JAMA* 288, no. 3 (2002): 334–41; C. F. Schairer et al., "Menopausal Estrogen and Estrogen-Progestin Replacement Therapy and Breast Cancer Risk," *JAMA* 283, no. 4 (2000): 485–91; F. Berrino et al., "Serum Sex Hormone Levels after Menopause and Subsequent Breast Cancer," *Journal of the National Cancer Institute* 88, no. 5 (1996): 291–96.

64. See, for example, S. Y. Mitchell et al., "Ovarian Hyperstimulation Syndrome Associated with Clomiphene Citrate," *West Indian Medical Journal* 50, no. 3 (2001): 227–29; A. Delvigne and S. Rozenberg, "Preventive Attitude of Physicians to Avoid OHSS in IVF Patients," *Human Reproduction* 16, no. 12 (2001): 2491–95; B. McElhinney and N. McClure, "Ovarian Hyperstimulation Syndrome," *Bailliere's Best Practice & Research: Clinical Obstetrics & Gynaecology* 14, no. 1 (2000): 103–22; J. G. Schenker, "Clinical Aspects of Ovarian Hyperstimulation Syndrome," *European Journal of Obstetric, Gynecology and Reproductive Biology* 85, no. 1 (1999): 13–20; H. S. Jacobs and R. Agrawal, "Complications of Ovarian Stimulation," *Bailliere's Clinical Obstetrics and Gynaecology* 12, no. 4 (1998): 565–79.

65. See, for example, L. Koch, "Physiological and Psychosocial Risks of the New Reproductive Technologies," in *Tough Choice*, ed. P. Stephenson and M. G. Wagner, 122–34 (Philadelphia: Temple University Press, 1993); P. J. Taylor and J. V. Kredentser, "Diagnostic and Therapeutic Laparoscopy and Hysteroscopy and Their Relationship to In Vitro Fertilization," in *A Textbook of In Vitro Fertilization and Assisted Reproductive Technology*, ed. P. R. Brinsden and P. A. Rainsbury, 73–92 (Park Ridge, NJ: The Parthenon Publishing Group, 1992).

66. American Society for Reproductive Medicine and Society for Assisted Reproductive Technology, "Assisted Reproductive Technology in the United States: 1999 Results Generated from the American Society for Reproductive Medicine/Society for Assisted Reproductive Technology Registry," *Fertility and Sterility* 78 (2002):

918–31; P. Lesny et al., "Transcervical Embryo Transfer as a Risk Factor for Ectopic Pregnancy," *Fertility and Sterility* 72 (1999): 305–9; A. Strandell, J. Thorburn, and L. Hamberger, "Risk Factors for Ectopic Pregnancy in Assisted Reproduction," *Fertility and Sterility* 71, no. 2 (1999): 282–86; S. F. Marcus and P. R. Brinsden, "Analysis of the Incidence and Risk Factors Associated with Ectopic Pregnancy Following In-Vitro Fertilization and Embryo Transfer," *Human Reproduction* 10, no. 1 (1995): 199–203.

Chapter Nine

On the Need to
Take Biology Seriously

Immanuel Kant once noted that the three most important questions for human beings are, "What can I know? What ought I to do? and What may I hope?" To these questions, some might add now, "What should I fear?" Kant argued that often our answers to these questions are confused, that we think we know things that we cannot, and based on these mistaken ideas we act in ways, and hope for things, that are unjustified or harmful to self or others.

Using modern science and technology has often been perceived as a way, if not the best way, to provide answers to these questions. In a world constantly shaped and reshaped by science and technology, many take great comfort in the notion that if we just can get the science right the answers are, or might be, clear and certain. These expectations are not foreign to modern biology. What could tell us more than a science of our very nature about what we can know, what we should do, or what we can reasonably hope or fear? And, what might be more exciting, interesting, and possibly horrifying than to know who we are, to make brilliantly clear our essence? Hence, according to an all-too-common misunderstanding of what contemporary biology can tell us and allow us to do, we might be able to control not only our very nature, but also our future, and the future of our children, in a way that will remove sources of suffering and fear.

Much of the excitement about the genetic revolution in biology in general, and the Human Genome Project in particular, can be understood in light of the questions Kant tried to answer. The promises of knowing who and what we are, of eliminating genetic diseases, of allowing infertile people to have genetically related children, of predicting and controlling dangerous or self-destructive behavior, and of giving our children the best possible future all inspire great hopes. Additionally, the knowledge we might gain about our capacities and limits might allow us to better allocate resources of time, money, and intellectual energy. As a result, we might improve not only ourselves as individuals, but also our societies. We could have less disease and crime; live longer, healthier lives; increase the number of people reaching their potential; and create more control over the course and content of our lives.

Conversely, many fear this knowledge and the future that this new world of genetic knowledge and control appears to promise. Cloning raises specters of fascist eugenics programs and fear of loss of human dignity. Genetic testing raises concerns about unfair discrimination. Others fear that if behavior is determined by unchangeable genetics, then notions of responsibility to others will be unsupportable, and resources will be further diverted from those most in need. All this leads to trepidation about an increasingly stratified and unjust society.

Given these hopes and fears, an evaluation of what biology can and cannot tell us about such issues seems not just important but necessary. In this book, I have argued that all too often these hopes and fears are ungrounded and that the presumed implications of genetics for ethics and social policy are unsupported. This book has then been a call to take biology seriously. To do so, of course, is not to use biology as a trump card. On the contrary, by carefully evaluating the role of biological and technological knowledge in ethical and public policy discussions, we can realize that biology cannot tell us many of the things we need to know about such issues. Taking biology seriously will prevent us from making claims about moral and public policy consequences that are grounded on epistemological, scientific, or ethical misunderstandings.

When we take biology seriously, we can realize that it is questionable to criticize genetic determinism by pointing out that if it is correct then individuals are not responsible for critically evaluating and maybe transforming inadequate institutions. Such criticisms, as we saw in chapter 2, commit an epistemological mistake. They simply misunderstand the role of biology in human life. And they do so because they ignore the fact that biological traits such as intelligence, aggression, addictive behavior, or differences in reproductive strategies between men and women can only be understood in the social context in which they appear. Such traits, as they are often defined in these discussions, cannot be said to be good or bad because of some intrinsic property they might have. If this is correct, then, as we have seen, a fear that social responsibility might be diminished is misplaced. A critical evaluation and transformation of our values and social arrangements appears to be more, not less, required were it the case that particular human traits and behaviors are genetically determined. Such evaluation can set those traits into the appropriate context, human's social environment, in order to judge the desirability or undesirability of such traits. Independently of this social context, the presence of these biological traits and behaviors cannot say much about social responsibility.

Similarly, when we take biology seriously, we can recognize that the debate over cloning human beings has often been framed in ways that misunderstand the biology behind this practice. As we saw in chapter 4, tying human dignity to the uniqueness of our genome is quite debatable because it is unclear how a biological entity such as our genome has anything to do with human dignity. Such a link is also dubious because of the difficulty of deciding what exactly it means to say that two genomes are identical or the same. In any case, even if we can agree on the fact that the genomes of a clone and its donor are relevantly similar, still it is difficult to see how this would interfere with human individuality. Identical twins have genomes that are more similar than the genomes of a clone and a nuclear donor need to be. Nevertheless, twins do have their own personalities, their own characters, and their own life choices. Paying attention to biology can remind us that human beings are very complex creatures influenced not just by our genes, but also by many other biological, environmental, and social factors. Likewise, carefully considering biological knowledge can inform us that cloning is not the tool to tackle genetic diseases. When we pay attention to biology, we can also appreciate that it cannot give us all the answers to moral and public policy issues related to human cloning. As we saw in chapter 5, even if it were the case that the scientific knowledge alleged to support the development and use of human cloning was correct, still this would not be a sufficient reason to support reproductive cloning. This would only be so if we presuppose not only that the biology is correct, but also that the social context in which cloning is developed and where claims about its moral adequacy are presented is irrelevant. We saw in that chapter that such an assumption is far from correct.

Finally, taking biology seriously can also tell us that many discussions about genetic technologies and genetic information are presented in ways that suggest that the predictive ability of genetic analysis is higher than actually is warranted. Examining this claim in chapter 7 brought to our attention that this mistake has made many proclaim that we have a moral obligation to obtain and share genetic information about ourselves. Nonetheless, although by paying careful attention to our current biological knowledge we can learn that a defense of duties related to our ability to obtain and share genetic information is grounded on a misunderstanding of human biology, biological knowledge by itself cannot tell us what our moral obligations regarding genetic information are. Thus, in chapter 8, I argued that even if it were the case that genetic testing was able to give us highly reliable information about future health status, this alone would not support claims about moral obligations to obtain and share genetic information. Attention to the social context in which these moral obligations would be binding to humans would also be necessary. Many of the debates on this issue pay, however, little attention to such context. Consequently, discussions on the subject of the alleged moral duties that follow from our ability to obtain genetic information about others and ourselves often ignore that access to these technologies is limited, that laypeople and professionals might not have adequate knowledge about genetics and genetic testing technologies, and that already disadvantaged groups might be unfairly burdened by these obligations.

This book has, then, been an attempt to call attention to the fact that good ethics requires good science, but it also calls for careful attention to the social context in which both the science and our ethical precepts and public policies play a role. It may well be that we are entering an era in which genetic science and technologies question our existing definitions of life and death, change our ideas about what a human being is, and transform the values we hold. And it is certainly the case that contemporary biology has much to contribute as we seek to know who we are, what we might know, how we ought to live, and what it is reasonable to hope and to fear. Taking biology seriously means paying careful attention to what it can, and what it cannot, tell us.

As noted throughout, epistemological, scientific, and ethical problems arise in much of our discussion of the implications of molecular genetics and genetic technologies because we fail to pay enough attention to the complexity of human biology and human life. There can be, of course, important heuristic reasons for this tendency. It is often easier to examine something in isolation. We do not, for instance, sequence genes while they are within cells. Still, the isolation is just that, a heuristic, and should not be taken to be the full story. Thus, whatever genetics might tell us about many human behaviors, we can only know the value and meanings of those behaviors when we place them in the actual social contexts within which human beings live. This includes recognizing that the social context might be malleable. We also cannot fully evaluate the possibilities of human cloning without examining in detail both the complexities of gene interactions and also our social institutions and social arrangements. Understanding the role of genes in disease and behavior further requires that we not think of genes in isolation, but within the larger biological and social processes of which they are a part. Finally, making sense of what duties we might have to obtain and share information about our genetic endowments will be an unsuccessful task if we fail to take account of both the complexity of human biology and the social context in which people really live and make decisions.

As we have seen throughout these pages, to neglect any of these aspects will likely misguide our efforts to improve our communities and better ourselves. Let us, then, take biology seriously. That is, let us pay attention to both what biology can and cannot tell us in this undertaking. We might then have more plausible ideas, better understand ourselves, avoid needless fears, and entertain more credible hopes.

Bibliography

Ahmadian, A., and J. Lundeberg. "A Brief History of Genetic Variation Analysis." *Biotechniques* 32, no. 5 (2002).

Alberts, B., et al. *Molecular Biology of the Cell*. 4th ed. New York: Garland Science, 2002.

American Society for Reproductive Medicine and Society for Assisted Reproductive Technology. "Assisted Reproductive Technology in the United States: 1999 Results Generated from the American Society for Reproductive Medicine/Society for Assisted Reproductive Technology Registry." *Fertility and Sterility* 78 (2002): 918–31.

Amundson, R. "Against Normal Function." *Studies in the History and Philosophy of Biological and Biomedical Sciences* 31, no. 1 (2000): 33–53.

Andre, J., L. M. Fleck, and T. Tomlinson. "On Being Genetically 'Irresponsible.'" *Kennedy Institute of Ethics Journal* 10, no. 2 (2000): 129–46.

Andrews, L. B. "Predicting and Punishing Antisocial Acts: How the Criminal Justice System Might Use Behavioral Genetics." In *Behavioral Genetics: The Clash of Culture and Biology*, edited by R. Carson and M. Rothstein, 116–55. Baltimore, MD: Johns Hopkins University Press, 1999.

Andrews, L. B., and M. Hibbert. "Courts and Wrongful Birth: Can Disability Itself Be Viewed as a Legal Wrong?" In *Americans with Disabilities: Exploring Implications of the Law for Individuals and Institutions*, edited by L. Pickering Francis and A. Silver, 318–30. New York: Routledge, 2000.

Ankeny, R. A. "Reduction Reconceptualized: Cystic Fibrosis as a Paradigm Case for Molecular Medicine." In *Mutating Concepts, Evolving Disciplines: Genetics, Medicine and Society*, edited by L. S. Parker and R. A. Ankeny, 127–41. Dordrecht, the Netherlands: Kluwer Academic Publishers, 2002.

Annas, G. "Scientific Discoveries and Cloning: Challenges for Public Policy." In *Flesh of My Flesh*, edited by G. Pence, 77–84. Lanham, MD: Rowman & Littlefield, 1998.

Arditti, R., R. D. Klein, and S. Minden. *Test-Tube Women*. London: Pandora Press, 1984.

Aristotle. *Nicomachean Ethics*. Translated by Terence Irwin. Indianapolis: Hackett, 1985.

Asch, A., and G. Geller. "Feminism, Bioethics, and Genetics." In *Feminism and Bioethics: Beyond Reproduction*, edited by S. Wolf. New York: Oxford University Press, 1996.

Badcock, C. *Evolutionary Psychology: A Critical Introduction*. Malden, MA: Polity Press, 2000.

Baier, A. *Moral Prejudices*. Cambridge, MA: Harvard University Press, 1994.

Balaban, E. "Behavior Genetics: Galen's Prophecy or Malpighi's Legacy?" In *Thinking about Evolution: Historical, Philosophical, and Political Perspectives*, vol. 2, edited by R. Singh et al., 429–66. Cambridge, UK: Cambridge University Press, 2001.

Baldano, J. L., and N. Katsanis. "Beyond Mendel: An Evolving View of Human Genetic Disease Transmission." *Nature Reviews: Genetics* 3, no. 10 (2002): 779–89.

Bankhead, C., et al. "New Developments in Genetics–Knowledge, Attitudes and Information Needs of Practice Nurses." *Family Practice* 18, no. 5 (2001): 475–86.

Barash, D. *The Whispering Within*. New York: Harper & Row, 1979.

Bekker, H., et al. "Uptake Of Cystic Fibrosis Testing in Primary Care: Supply Push or Demand Pull?" *British Medical Journal* 306 (1993): 1584–86.

Benton, T. "Social Causes and Natural Relations." In *Alas, Poor Darwin: Arguments against Evolutionary Psychology*, edited by H. Rose and S. Rose, 249–72. New York: Harmony Books, 2000.

Berrino, F., et al. "Serum Sex Hormone Levels after Menopause and Subsequent Breast Cancer." *Journal of the National Cancer Institute* 88, no. 5 (1996): 291–96.

Bertram, L., and R. E. Tanzi. "Alzheimer's Disease: One Disorder, Too Many Genes?" *Human Molecular Genetics* 13, spec. no. 1: (2004): R135–R141.

Betts, D., et al. "Reprogramming of Telomerase Activity and Rebuilding of Telomere Length in Cloned Cattle." *Proceedings of the National Academy of Sciences of USA* 98, no. 3 (2001): 1077–82.

Beurton, P. J., H.-J. Rheinberger, and R. Falk, eds. *The Concept of the Gene in Development and Evolution*. Cambridge, UK: Cambridge University Press, 2000.

Bird, T. D., et al. "Wide Range in Age of Onset for Chromosome 1–Related Familial Alzheimer Disease." *Annals of Neurology* 40, no. 6 (1996): 932–36.

Bleier, R. *Science and Gender: A Critique of Biology and Its Theories on Women*. New York: Pergamon Press, 1985.

Botkin, J. "Ethical Issues and Practical Problems in Preimplantation Genetic Diagnosis." *Journal of Law, Medicine & Ethics* 26, no. 1 (1998): 17–28.

Bottorff, J. L., et al. "Communicating Cancer Risk Information: The Challenges of Uncertainty." *Patient Education and Counseling* 33, no. 1 (1998): 67–81.

Breslow, R. A., et al. "Americans' Knowledge of Cancer Risk and Survival." *Preventive Medicine* 26, no. 2 (1997): 170–77.

Bristow, R. E., and B. Y. Karlan. "The Risk of Ovarian Cancer after Treatment for Infertility." *Current Opinion in Obstetric and Gynecology* 8, no. 1 (1996): 32–37.

Brock, D. "Cloning Human Beings: An Assessment of the Ethical Issues Pro and Con." In *Clones and Clones*, edited by M. C. Nussbaum and C. R. Sunstein, 153–54. New York: W. W. Norton & Company, 1998.

Brooks, A. "Women's Voices: Prenatal Diagnosis and Care for the Disabled." *Health Care Analysis* 9 (2001): 133–50.

Buchanan, A. "Choosing Who Will Be Disabled: Genetic Intervention and the Morality of Inclusion." *Social Philosophy and Policy* 13, no. 1 (1996): 18–46.

Buchanan, A., D. Brock, N. Daniels, and D. Wikler. *From Chance to Choice: Genetics and Justice.* Cambridge, UK: Cambridge University Press, 2000.

Burke, W., and J. Emery. "Genetic Education for Primary-Care Providers." *Nature Reviews: Genetics* 3, no. 7 (2002): 561–66.

Burnham, T., and J. Phelan. *Mean Genes.* Cambridge, MA: Perseus Publishing, 2000.

Buss, D. *Evolutionary Psychology: The New Science of the Mind.* Boston: Allyn & Bacon, 1999.

Buss, D. "Sexual Conflict: Evolutionary Insights into Feminism and the 'Battle of the Sexes.'" In *Sex, Power, Conflict: Evolutionary and Feminist Perspectives*, edited by D. M. Buss and N. M. Malamuth, 296–318. New York: Oxford University Press, 1996.

Butler, D. "The Fertility Riddle." *Nature* 432, no. 7013 (2004): 38–39.

Callahan, D. "Perspective on Cloning: A Threat to Individual Uniqueness." *Los Angeles Times*, November 12, 1993, B7.

Callahan, J. C., ed. *Reproduction, Ethics, and The Law.* Bloomington, IN: Indiana University Press, 1995.

Carlson, L. "The Morality of Prenatal Testing and Selective Abortion: Clarifying the Expressivist Objection." In *Mutating Concepts, Evolving Disciplines: Genetics, Medicine and Society*, edited by L. S. Parker and R. A. Ankeny, 191–213. Dordrecht, the Netherlands: Kluwer Academic Publishers, 2002.

Carse, A. L. "Impartial Principle and Moral Context: Securing a Place for the Particular in Ethical Theory." *Journal of Medicine and Philosophy* 23, no. 2 (1998): 153–69.

Cartwright, J. *Evolution and Human Behavior.* Cambridge, MA: The MIT Press, 2000.

Caulfield, T. "Human Cloning Laws, Human Dignity and the Poverty of the Policy Making Dialogue." *BMC Medical Ethics* 4, no. 1 (2003).

Chinnery, P. F., et al. "Accumulation of Mitochondrial DNA Mutations in Ageing, Cancer, and Mitochondrial Disease: Is There a Common Mechanism?" *Lancet* 360, no. 9342 (2002): 1323–25.

Clarkeburn, H. "Parental Duties and Untreatable Genetic Conditions." *Journal of Medical Ethics* 26, no. 5 (2000): 400–405.

Collins, F. *A Brief Primer on Genetic Testing.* National Human Genome Research Institute, 2003. Available at http://www.genome.gov/10506784 (accessed 16 Feb. 2005).

———. "BRCA1—Lots of Mutations, Lots of Dilemmas." *New England Journal of Medicine* 334 (1996): 186–88.

Collins, F., and V. McKusick. "Implications of the Human Genome Project for Medical Science." *JAMA* 285, no. 5 (2001): 540–44.

Comings, D. E., et al. "The Additive Effect of Neurotransmitter Genes in Pathological Gambling." *Clinical Genetics* 60, no. 2 (2001): 107–16.

Comisión Especial de Estudio de la Fecundación "In Vitro" y la Inseminación Artificial Humanas [Special Commission for the Study of Human In Vitro Fertilization and Artificial Insemination]. *Informe [Report]*. Madrid, Spain: Gabinete de Publicaciones, 1987.

Committee on Science, Engineering, and Public Policy and Global Affairs Division. *Scientific and Medical Aspects of Human Reproductive Cloning*. Washington, DC: National Academy Press, 2002.

Coughlin, M. J. "Ethical Implications of Genetic Testing for Breast Cancer Susceptibility." *Critical Reviews in Oncology/Hematology* 40, no. 2 (2001): 149–57.

Council of Europe. *Additional Protocol to the Convention for the Protection of Human Rights and Dignity of the Human Being with Regard to the Application of Biology and Medicine, on the Prohibition of Cloning Human Beings*. Paris, France: Council of Europe, 1998. Available at http://www.virtual-institute.de/en/hp/ embryo/regional/AP12011998.pdf (accessed 16 Jan. 2005).

Cranor, C. "Genetic Causation." In *Are Genes Us? The Social Consequences of the New Genetics*, edited by C. Cranor, 125–41. New Brunswick, NJ: Rutgers University Press, 1994.

Cummings, M. *Human Heredity: Principles and Issues*. 5th ed. Pacific Grove, CA: Brooks/Cole, 2000.

d'Agincourt-Canning, L. "Experiences of Genetic Risk: Disclosure and the Gendering of Responsibility." *Bioethics* 15, no. 3 (2001): 231–47.

Daly, M., and M. Wilson. "Evolutionary Psychology and Marital Conflict." In *Sex, Power, Conflict: Evolutionary and Feminists Perspectives*, edited by D. M. Buss and N. M. Malamuth, 9–28. New York: Oxford University Press, 1996.

———. *Sex, Evolution, and Behavior*. Boston: Willard Grant, 1983.

Davis, D. "Genetic Dilemmas and the Child's Right to an Open Future." *The Hastings Center Report* 27, no. 2 (1997): 7–15.

Dawkins, R. *The Extended Phenotype: The Long Reach of the Gene*. New York: W. H. Freeman, 1982.

———. *The Selfish Gene*. 2nd ed. New York: Oxford University Press, 1989.

Decruyenaere, M., etal. "Cystic Fibrosis: Community Knowledge and Attitudes Towards Carrier Screening and Prenatal Diagnosis," *Clinical Genetics* 41 (1992): 189–96.

Delvigne, A., and S. Rozenberg. "Preventive Attitude of Physicians to Avoid OHSS in IVF Patients." *Human Reproduction* 16, no. 12 (2001): 2491–95.

de Melo-Martín, I. *Making Babies: Biomedical Technologies, Reproductive Ethics, and Public Policy*. Dordrecht, the Netherlands: Kluwer, 1998.

de Melo-Martín, I., and C. Hanks. "Genetic Technologies and Women: The Importance of Context." *Bulletin of Science, Technology & Society* 21, no. 5 (2001): 354–60.

DeNavas-Walt, C., B. D. Proctor, and R. J. Mills, U.S. Census Bureau, Current Population Reports, P60-226. *Income, Poverty, and Health Insurance Coverage in the United States: 2003*. Washington, DC: U.S. Government Printing Office, 2004. Available at http://www.census.gov/prod/2004pubs/p60-226.pdf (accessed 21 Feb. 2005).

Dennis, C. "Epigenetics and Disease: Altered States." *Nature* 421, no. 6924 (2003): 686–89.

Devlin, B., S. Fienberg, D. Resnick, and K. Roeder, eds. *Intelligence, Genes, and Success: Scientists Respond to The Bell Curve*. New York: Springer-Verlag, 1997.

Di Berardino, M. A., and R. G. McKinnell. "The Pathway to Animal Cloning and Beyond—Robert Briggs (1911–1983) and Thomas J. King (1921–2000)." *Journal of Experimental Zoology: Part A, Comparative Experimental Biology* 301, no. 4 (2004): 275–79.

Dillon, S. "Harvard Chief Defends His Talk on Women." *New York Times*, January 18, 2005.

DiMauro, S., and E. A. Schon. "Mitochondrial DNA Mutations in Human Disease." *American Journal of Medical Genetics* 106, no. 1 (2001): 18–26.

Dipple, K. M., and E. McCabe. "Modifier Genes Convert 'Simple' Mendelian Disorders to Complex Traits." *Molecular Genetics and Metabolism* 71, nos. 1–2 (2000): 43–50.

Diver, C., and J. Cohen. "Genophobia: What Is Wrong with Genetic Discrimination?" *University of Pennsylvania Law Review* 149, no. 5 (2001): 1439–82.

Dolan, N. C., A. M. Lee, and M. M. McDermott. "Age-Related Differences in Breast Carcinoma Knowledge, Beliefs, and Perceived Risk among Women Visiting an Academic General Medicine Practice." *Cancer* 80, no. 3 (1997): 413–20.

Donelan, K., M. Falik, and C. M. DesRoches. "Caregiving: Challenges and Implications for Women's Health." *Women's Health Issues* 11, no. 3 (2001): 185–200.

Duaux, E., M. O. Krebs, H. Loo, and M. F. Poirier. "Genetic Vulnerability To Drug Abuse." *European Psychiatry* 15, no. 2 (2000): 9–14.

Dupré, J. *Human Nature and the Limits of Science*. New York: Oxford University Press, 2001.

——. *The Disorder of Things: Metaphysical Foundations of the Disunity of Science*. Cambridge, MA: Harvard University Press, 1995.

Durr, A., et al. "Diagnosis of 'Sporadic' Huntington's Disease." *Journal of the Neurological Sciences* 129, no. 1 (1995): 51–55.

Ellis, G. B. "Infertility and the Role of the Federal Government." In *Beyond Baby M*, edited by D. M. Bartels et al., 111–30. Clifton, NJ: Humana Press, 1990.

Emery, J., and S. Hayflick. "The Challenge of Integrating Genetic Medicine into Primary Care." *British Medical Journal* 322, no. 7293 (2001): 1027–30.

Emery, J., et al. "A Systematic Review of the Literature Exploring the Role of Primary Care in Genetic Services." *Family Practice* 16, no. 4 (1999): 426–45.

Evers-Kiebooms, G., et al. "A Stigmatizing Effect of the Carrier Status for Cystic Fibrosis?" *Clinical Genetics* 46, no. 5 (1994): 336–43.

Farrer, L. A., et al. "Effects of Age, Sex, and Ethnicity on the Association between Apolipoprotein E Genotype and Alzheimer Disease: A Meta-Analysis." *JAMA* 278, no. 16 (1997): 1349–56.

Fausto-Sterling, A. *Myths of Gender*. New York: Basic Books, 1985.

——. *Sexing the Body: Gender Politics and the Construction of Sexuality*. New York: Basic Books, 2000.

Feinberg, J. "The Child's Right to an Open Future." In *Whose Child? Children's Rights, Parental Authority, and State Power*, edited by W. Aiken and H. LaFollette, 124–53. Totowa, NJ: Rowman & Littlefield, 1980.

Feldman, D. "Human Dignity as a Legal Value: Part 1." *Public Law* 4 (Winter 1999): 682–702.

Fetters, M. D., D. J. Doukas, and K. L. D. Phan. "Family Physicians' Perspectives on Genetics and the Human Genome Project." *Clinical Genetics* 56 (1999): 28–34.

Fitzgerald, J. "Geneticizing Disability: The Human Genome Project and the Commodification of Self." *Issues in Law & Medicine* 14, no. 2 (1998): 147–63.

Fong, K. M., et al. "Lung Cancer • 9: Molecular Biology of Lung Cancer: Clinical Implications." *Thorax* 58, no. 10 (2003): 892–900.

Foot, P. *Virtues and Vices*. Oxford, UK: Blackwell, 1978.

Ford, D., et al. "Genetic Heterogeneity and Penetrance Analysis of the BRCA1 and BRCA2 Genes in Breast Cancer Families." *American Journal of Human Genetics* 62, no. 3 (1998): 676–89.

Frankena, W. K. "The Naturalistic Fallacy." *Mind* 48, no. 192 (1939): 464–77.

Fraser, S., ed. *The Bell Curve Wars: Race, Intelligence, and the Future of America*. New York: Basic Books, 1995.

Fuerst, J. A. "The Role of Reductionism in the Development of Molecular Biology: Peripheral or Central?" *Social Studies of Science* 12, no. 2 (1982): 241–78.

Fukuyama, F. *Our Posthuman Future: Consequences of the Biotechnology Revolution*. New York: Farrar, Straus and Giroux, 2002.

Galton, D. J., and G. A. Ferns. "Genetic Markers to Predict Polygenic Disease: A New Problem for Social Genetics." *QJM* 92, no. 4 (1999): 223–32.

Gannett, L. "What's in a Cause? The Pragmatic Dimensions of Genetic Explanations." *Biology and Philosophy* 14, no. 3 (1999): 349–74.

———. "Tractable Genes, Entrenched Social Structures." *Biology and Philosophy* 12, no. 3 (1997): 403–19.

Garber, J. "A 40-Year-Old Woman with a Strong Family History of Breast Cancer." *JAMA* 282, no. 20 (1999): 1953.

Garinis, G.A., et al. "DNA Hypermethylation: When Tumor Suppressor Genes Go Silent." *Human Genetics* 111, no. 2 (2002): 115–27.

Genetics and Public Policy Center. *Genetic Testing*. Genetics and Public Policy Center, 2005. Available at http://www.dnapolicy.org/genetics/testing.jhtml (accessed 21 Feb. 2005).

———. *Preimplantation Genetic Diagnosis*. Genetics and Public Policy Center, 2004. Available at http://www.dnapolicy.org/downloads/pdfs/policy_pgd.pdf (accessed 31 Jan. 2005).

Gewirth, A. "Human Dignity as the Basis of Rights." In *The Constitution of Rights: Human Dignity and American Values*, edited by M. Meyer and W. Parent, 10–46. London: Cornell University Press, 1992.

Gibbs, W. W. "The Unseen Genome: Beyond DNA." *Scientific American* 289, no. 6 (2003): 106–13.

Gifford, F. "Understanding Genetic Causation and Its Implications for Ethical Issues in Human Genetics." In *Mutating Concepts, Evolving Disciplines: Genetics, Medicine, and Society*, edited by R. Ankeny and L. Parker, 109–25. Dordrecht, the Netherlands: Kluwer Academic Publishers, 2002.

Gilbert, W. "A Vision of the Grail." In *The Code of Codes: Scientific and Social Issues in the Human Genome Project*, edited by D. Kevles and L. Hood, 83–97. Cambridge, MA: Harvard University Press, 1992.

Gilligan, C. *In a Different Voice: Psychological Theory and Women's Development.* Cambridge, MA: Harvard University Press, 1982.

Glannon, W. *Genes and Future People.* Boulder, CO: Westview Press, 2001.

Glover, J. *What Sort of People Should There Be?* New York: Penguin, 1984.

Goelen, G., et al. "Moral Concerns of Different Types of Patients in Clinical BRCA1/2 Gene Mutation Testing." *Journal of Clinical Oncology* 17, no. 5 (1999): 1595–1600.

Gordon, C., et al. "Population Screening for Cystic Fibrosis: Knowledge and Emotional Consequences 18 Months Later." *American Journal of Medical Genetics* 120, no. 2 (2003): 199–208.

Gotz, M. J., E. C. Johnstone, and S. G. Ratcliffe. "Criminality and Antisocial Behaviour in Unselected Men with Sex Chromosome Abnormalities." *Psychological Medicine*, 29, no. 4 (1999): 953–62.

Gould, S. J. *The Mismeasure of Man.* New York: W. W. Norton, 1981.

Green, R. C. "Risk Assessment for Alzheimer's Disease with Genetic Susceptibility Testing: Has the Moment Arrived?" *Alzheimer's Care Quarterly* 3, no. 3 (2002): 208–14.

Griffiths, P., and E. Neumann-Held. "The Many Faces of the Gene." *BioScience* 49, no. 8 (1999): 656–62.

Grunfeld, E. A., et al. "Women's Knowledge and Beliefs Regarding Breast Cancer." *British Journal of Cancer* 86 (2002): 1373–78.

Gurdon, J. B., and J. A. Byrne. "The First Half-Century of Nuclear Transplantation." *Proceedings from the National Academy of Science of USA* 100, no. 14 (2003): 8048–52.

Habermas, J. *The Future of Human Nature.* Cambridge, UK: Polity, 2003.

Hallowell, N. "Doing the Right Thing: Genetic Risk and Responsibility." *Sociology of Health and Illness* 21, no. 5 (1999): 597–621.

Hallowell, N., H. Statham, and F. Murton. "Women's Understanding of Their Risk of Developing Breast/Ovarian Cancer Before and After Genetic Counseling." *Journal of Genetic Counseling* 7, no. 4 (1998): 345–64.

Hallowell, N., et al. "Balancing Autonomy and Responsibility: The Ethics of Generating and Disclosing Genetic Information." *Journal of Medical Ethics* 29, no. 2 (2003): 74–79.

Harris, J. *The Value of Life.* London: Routledge, 1985.

Held, V. *Feminist Morality.* Chicago: University of Chicago Press, 1993.

Hellman, D. "What Makes Genetic Discrimination Exceptional?" *American Journal of Law and Medicine* 29, no. 1 (2003): 77–116.

Henifin, M. S. "New Reproductive Technologies: Equity and Access to Reproductive Health." *Journal of Social Issues* 49, no. 2 (1993): 61–74.

Herissone-Kelly, P. "Bioethics in the United Kingdom: Genetic Screening, Disability Rights, and the Erosion of Trust." *Cambridge Quarterly of Healthcare Ethics* 12, no. 3 (2003): 235–41.

Herrnstein, R., and C. Murray. *The Bell Curve: Intelligence and Class Structure in American Life*. New York: Free Press, 1994.

Ho, L., et al. "The Molecular Biology of Huntington's Disease." *Psychological Medicine* 31, no. 1 (2001): 3–14.

Hoffmann, R. L., and A. M. Mitchell. "Caregiver Burden: Historical Development." *Nursing Forum* 33, no. 4 (1998): 5–11.

Holzmann, C., et al. "Functional Characterization of the Human Huntington's Disease Gene Promoter." *Molecular Brain Research* 92, nos. 1–2 (2001): 85–97.

Honda, K., and A. I. Neugut. "Associations between Perceived Cancer Risk and Established Risk Factors in a National Community Sample." *Cancer Detection and Prevention* 28, no. 1 (2004): 1–7.

Hopper, J. L., et al. "Population-Based Estimate of the Average Age-Specific Cumulative Risk of Breast Cancer for a Defined Set of Protein-Truncating Mutations in BRCA1 and BRCA2: Australian Breast Cancer Family Study." *Cancer Epidemiology, Biomarkers & Prevention* 8, no. 9 (1999): 741–47.

Hubbard, R. "Genetics and Women's Health." *Journal of the American Medical Women's Association* 52, no. 1 (1997): 2–3.

Hubbard, R., and E. Wald. *Exploding the Gene Myth*. Boston, MA: Beacon Press, 1997.

Hudson, K., et al. "Genetic Discrimination and Health Insurance: An Urgent Need for Reform." *Science* 270, no. 5235 (1995): 391–93.

Hughes, W. "Richard's Defense of Evolutionary Ethics." *Biology and Philosophy* 1 (1986): 306–15.

Huibers, A., and A. van 't Spijker. "The Autonomy Paradox: Predictive Genetic Testing and Autonomy; Three Essential Problems." *Patient Education and Counseling* 35, no. 1 (1998): 53–62.

Hume, D. *Treatise of Human Nature*. Oxford, UK: Oxford University Press, 1968.

Humpherys, D., et al. "Abnormal Gene Expression in Cloned Mice Derived from Embryonic Stem Cell and Cumulus Cell Nuclei." *Proceedings of the National Academy of Sciences* 99, no. 20 (2002): 12889–94.

Huntington's Disease Collaborative Research Group. "A Novel Gene Containing a Trinucleotide Repeat That Is Expanded and Unstable on Huntington's Disease Chromosomes." *Cell* 72, no. 6 (1993): 971–83.

Hwang, W. S., et al. "Evidence of a Pluripotent Human Embryonic Stem Cell Line Derived from a Cloned Blastocyst." *Science* 303, no. 5664 (2004): 1669–74.

Illmensee, K. "Cloning in Reproductive Medicine." *Journal of Assisted Reproduction and Genetics* 18, no. 8 (2001): 451–67.

Institute of Medical Genetics. *Frequency of Inherited Disorders Database*. University of Wales College of Medicine. Available at http://archive.uwcm.ac.uk/uwcm/mg/fidd/ (accessed 21 Feb. 2005).

Ireland, M. S. *Reconceiving Women*. New York: Guilford Press, 1993.

Irvine, D. S. "Male Reproductive Health: Cause for Concern?" *Andrologia* 32, nos. 4–5 (2000): 195–208.

Jackson, R. *Mothers Who Leave*. London, UK: Pandora, 1994.

Jacobs, H. S., and R. Agrawal. "Complications of Ovarian Stimulation." *Bailliere's Clinical Obstetrics and Gynaecology* 12, no. 4 (1998): 565–79.

Jacoby, R., and N. Glauberman, ed. *The Bell Curve Debate*. New York: Times Books, 1995.

Jaenisch, R., and A. Bird. "Epigenetic Regulation of Gene Expression: How the Genome Integrates Intrinsic and Environmental Signals." *Nature Genetics* 33, suppl. (2003): 245–54.

Jaenisch, R., et al. "Nuclear Cloning, Stem Cells, and Genomic Reprogramming." *Cloning Stem Cells* 4, no. 4 (2002): 389–96.

Jarvik, L. F., V. Klodin, and S. S. Matsuyama. "Human Aggression and the Extra Y Chromosome: Fact or Fantasy?" *American Psychologist* 28, no. 8 (1973): 674–82.

Jeffery, C. R. *Criminology: An Interdisciplinary Approach*. Englewood Cliffs, NJ: Prentice Hall, 1990.

Jensen, A. *The g Factor: The Science of Mental Ability*. Westport, CT: Praeger Publishers, 1998.

Jensen, T. K., et al. "Poor Semen Quality May Contribute to Recent Decline in Fertility Rates." *Human Reproduction* 17, no. 6 (2002): 1437–40.

Jewelewicz, R., and E. E. Wallach. "Evaluation of the Infertile Couple." In *Reproductive Medicine and Surgery*, edited by E. E. Wallach and H. A. Zacur, 363–75. St. Louis, MO: Mosby, 1994.

Johnson, A. *Genetics and Health Insurance*. Denver, CO: NCSL, 2002. Available at http://www.ncsl.org/programs/health/genetics/Geneticshealthins.pdf (accessed 14 Feb. 2005).

Johnson, B. B., and P. Slovic. "Presenting Uncertainty in Health Risk Assessment: Initial Studies of Its Effects on Risk Perception and Trust." *Risk Analysis* 15, no. 4 (1995): 485–94.

Jonas, H. *Philosophical Essays: From Ancient Creed to Technological Man*. Englewood Cliffs, NJ: Prentice Hall, 1974.

Jones, P. A., and P. W. Laird. "Cancer Epigenetics Comes of Age." *Nature Genetics* 21, no. 2 (1999): 163–67.

Jonsen, A., and S. Toulmin. *The Abuse of Casuistry*. Berkeley, CA: University of California Press, 1988.

Juengst, E. "Can Enhancement Be Distinguished from Prevention in Genetic Medicine?" *Journal of Medicine and Philosophy* 22, no. 2 (1997): 125–42.

Kaback, M. "Population-Based Genetic Screening for Reproductive Counseling: The Tay-Sachs Disease Model." *European Journal of Pediatrics* 159, suppl. 3 (2000): S192–95.

Kahn, J. "Ethical Issues in Genetic Testing for Alzheimer's Disease." *Geriatrics* 52, no. 9 (1997): 30–32.

Kahneman, D., P. Slovic, and A. Tversky, ed. *Judgment Under Uncertainty: Heuristics and Biases*. Cambridge, UK: Cambridge University Press, 1974.

Kalebic, T. "Epigenetic Changes: Potential Therapeutic Targets." *Annals of the New York Academy of Sciences* 983, no. 1 (2003): 278–85.

Kamboh, M. I. "Molecular Genetics of Late-Onset Alzheimer's Disease." *Annals of Human Genetics* 68, pt. 4 (2004): 381–404.

Kaplan, J. *The Limits and Lies of Human Genetic Research*. New York: Routledge, 2000.

Kaplan, J., and M. Pigliucci. "Genes 'for' Phenotypes: A Modern History View." *Biology and Philosophy* 16, no. 2 (2001): 189–213.

Kass, L. "The Wisdom of Repugnance." In *Flesh of My Flesh*, edited by G. Pence, 13–37. Lanham, MD: Rowman & Littlefield, 1998.

Kevles, D. *In the Name of Eugenics: Genetics and the Uses of Human Heredity*. New York: Knopf, 1995.

Kevles, D., and L. Hood. *The Code of Codes*. Cambridge, MA: Harvard University Press, 1992.

Khorasanizadeh, S. "The Nucleosome: From Genomic Organization to Genomic Regulation." *Cell* 116, no. 2 (2004): 259–72.

Khoury, S. S., and K. K. Steinberg. "BRCA1 and BRCA2 Gene Mutations and Risk of Breast Cancer: Public Health Perspectives." *American Journal of Preventive Medicine* 16, no. 2 (1999): 91–98.

Kinmonth, A. L., J. Reinhard, and S. Pauker. "The New Genetics: Implications for Clinical Service in Britain and the United States." *British Medical Journal* 316, no. 7133 (1998): 767–70.

Kitcher, P. *The Lives to Come: The Genetic Revolution and Human Possibilities*. New York: Simon & Schuster, 1997.

——. "1953 and All That: A Tale of Two Sciences." *Philosophical Review* 93, no. 3 (1984): 335–76.

——. *Vaulting Ambition: Sociobiology and the Quest for Human Nature*. Cambridge, MA: The MIT Press, 1985.

——. "Whose Self Is It, Anyway?" *The Sciences* 37, no. 5 (1997): 58–62.

Kittay, E. F. *Love's Labor*. New York: Routledge, 1999.

Koch, L. "Physiological and Psychosocial Risks of the New Reproductive Technologies." In *Tough Choices*, edited by P. Stephenson and M. G. Wagner, 122–34. Philadelphia: Temple University Press, 1993.

Koch, T. "Disability and Difference: Balancing Social and Physical Constructions." *Journal of Medical Ethics* 27, no. 6 (2001): 370–76.

Kolonel, L. N., D. Altshuler, and B. E. Henderson. "The Multiethnic Cohort Study: Exploring Genes, Lifestyle and Cancer Risk." *Nature Reviews: Cancer* 4, no. 7 (2004): 519–27.

Koval, R., and J. A. Scutt. "Genetic and Reproductive Engineering—All for the Infertile?" In *Baby Machine*, edited by J. A. Scutt, 33–57. Melbourne, Australia: McCulloch Publishing, 1988.

Lacey, J. V., Jr., et al. "Menopausal Hormone Replacement Therapy and Risk Of Ovarian Cancer." *JAMA* 288, no. 3 (2002): 334–41.

Lawrence, W. F., et al. "Cost of Genetic Counseling and Testing for BRCA1 and BRCA2 Breast Cancer Susceptibility Mutations." *Cancer Epidemiology Biomarkers & Prevention* 10, no. 5 (2001): 475–81.

Lee, R. S., et al. "Cloned Cattle Fetuses with the Same Nuclear Genetics Are More Variable Than Contemporary Half-Siblings Resulting from Artificial Insemination and Exhibit Fetal and Placental Growth Deregulation Even in the First Trimester." *Biology of Reproduction* 70, no. 1 (2004): 1–11.

Lee, T. *The Human Genome Project: Cracking the Genetic Code of Life*. New York: Plenum, 1991.

Lemmens, T. "Selective Justice, Genetic Discrimination, and Insurance: Should We Single Out Genes in Our Laws?" *McGill Law Journal* 45, no. 2 (2000): 347–12.

Lesny, P., et al. "Transcervical Embryo Transfer as a Risk Factor for Ectopic Pregnancy." *Fertility and Sterility* 72 (1999): 305–9.

Levy-Lahad, E., and S. Plon. "A Risky Business: Assessing Breast Cancer Risk." *Science* 302, no. 5645 (2003): 574–75.

Lewontin, R. *The Triple Helix*. Cambridge, MA: Harvard University Press, 2000.

Lippman, A. "Prenatal Genetic Testing and Screening: Constructing Needs and Reinforcing Inequities." *American Journal of Law and Medicine* 17, nos. 1–2 (1991): 15–50.

———. "Worrying—and Worrying about—the Geneticization of Reproduction and Health." In *Misconceptions: The Social Construction of Choice and the New Reproductive Technologies*, vol. 1, edited by G. Basen, M. Eichler, and A. Lippman, 39–65. Ottawa, Canada: Voyageur Press, 1993.

Lippman-Hand, A., and F. C. Fraser. "Genetic Counseling: Provision and Reception of Information." *American Journal of Medical Genetics* 3, no. 2 (1979): 113–27.

Longino, H. "Behavior as Affliction: Common Frameworks of Behavior Genetics and Its Rivals." In *Mutating Concepts, Evolving Disciplines: Genetics, Medicine, and Society*, edited by R. Ankeny and L. Parker, 165–87. Dordrecht, the Netherlands: Kluwer Academic Publishers, 2002.

Lorentz, C. P., et al. "Primer on Medical Genomics Part I: History of Genetics and Sequencing of the Human Genome." *Mayo Clinic Proceedings* 77, no. 8 (2002): 773–82.

Luch, A. "Nature and Nurture—Lessons from Chemical Carcinogenesis." *Nature Reviews: Cancer* 5, no. 2 (2005): 113–25.

Macgregor, R. B., and G. M. Poon. "The DNA Double Helix Fifty Years On." *Computational Biology and Chemistry* 27, nos. 4–5 (2003): 461–67.

MacIntyre, A. *After Virtue*. Notre Dame, IN: University of Notre Dame Press, 1981.

———. *Dependent Rational Animals: Why Human Beings Need the Virtues*. Chicago: Open Court, 1999.

———. "Hume on 'Is' and 'Ought.'" *Philosophical Review* 68, no. 4 (1959): 451–68.

Macking, R. "Splitting Embryos on the Slippery Slope: Ethics and Public Policy." *Kennedy Institute of Ethics Journal* 4 (1994): 209–26.

Mahowald, M. "A Feminist Standpoint for Genetics." *Journal of Clinical Ethics* 7, no. 4 (1996): 333–40.

———. *Genes, Women, Equality*. New York: Oxford University Press, 2000.

Mahowald, M., et al. *Genetics in the Clinic: Clinical, Ethical, and Social Implications for Primary Care*. Philadelphia: Mosby, 2001.

Manuel, C. E. "Physician-Assisted Suicide Permits Dignity in Dying: Oregon Takes on Attorney General Ashcroft." *The Journal of Legal Medicine* 23, no. 4 (2002): 563–86.

Marcus, S. F., and P. R. Brinsden. "Analysis of the Incidence and Risk Factors Associated with Ectopic Pregnancy Following In-Vitro Fertilization and Embryo Transfer." *Human Reproduction* 10, no. 1 (1995): 199–203.

Marteau, T. M., et al. "Long-Term Cognitive and Emotional Impact of Genetic Testing for Carriers of Cystic Fibrosis: The Effects of Test Result and Gender." *Health Psychology* 16, no. 1 (1997): 51–62.

Mason, J. "Gender, Care and Sensibility in Family and Kin Relationships." In *Sex, Sensibility and the Gendered Body*, edited by J. Holland and L. Adkins, 15–36. Basingstoke, UK: Macmillan, 1996.

McElhinney, B., and N. McClure. "Ovarian Hyperstimulation Syndrome." *Bailliere's Best Practice & Research: Clinical Obstetrics & Gynaecology* 14, no. 1 (2000): 103–22.

Mealey, L. *Sex Differences: Developmental and Evolutionary Strategies*. San Diego: Academic Press, 2000.

Meigs, C. *Women and Their Diseases*. Philadelphia: Lea & Blanchard, 1848.

Metcalfe, S., et al. "Needs Assessment Study of Genetics Education for General Practitioners in Australia." *Genetics in Medicine* 4, no. 2 (2002): 71–77.

Mickle, J. E., and G. R. Cutting. "Genotype-Phenotype Relationships in Cystic Fibrosis." *The Medical Clinics of North America* 84, no. 3 (2000): 597–607.

Midgley, M. *Beast and Man: The Roots of Human Nature*. London: Routledge, 1996.

Miller, P. S. "Genetic Discrimination in the Workplace." *Journal of Law, Medicine and Ethics* 26, no. 3 (1998): 189–97.

Mitchell, S. Y., et al. "Ovarian Hyperstimulation Syndrome Associated with Clomiphene Citrate." *West Indian Medical Journal* 50, no. 3 (2001): 227–29.

Mollard, R., M. Denham, and A. Trounson. "Technical Advances and Pitfalls on the Way to Human Cloning." *Differentiation* 70, no. 1 (2002): 1–9.

Moore, G. E. *Principia Ethica*. Cambridge, UK: Cambridge University Press, 1903.

Moss, L. "Deconstructing the Gene and Reconstructing Molecular Developmental Systems." In *Cycles of Contingency: Developmental Systems and Evolution*, edited by S. Oyama, P. Griffiths, and R. Gray, 85–97. Cambridge, MA: The MIT Press, 2001.

———. "Deconstructing the Gene and Reconstructing Molecular Developmental Systems." In *Cycles of Contingency: Developmental Systems and Evolution*, edited by S. Oyama, P. Griffiths, and R. Gray, 85–97. Cambridge, MA: The MIT Press, 2001.

———. *What Genes Can't Do*. Cambridge, MA: The MIT Press, 2002.

Mueller, B. A., and J. R. Daling. "The Epidemiology of Infertility." In *Controversies in Reproductive Endocrinology and Infertility*, edited by M. R. Soules, 1–13. New York: Elsevier, 1989.

Munévar, G. "The Morality of Rational Ants." In *The Naked Truth: A Darwinian Approach to Philosophy*, 131–47. Aldershot, UK: Ashgate, 1998.

Murphy, J. S. "Should Lesbians Count as Infertile Couples? Antilesbian Discrimination in Assisted Reproduction." In *Embodying Bioethics*, edited by A. Donchin and L. Purdy, 103–20. Lanham, MD: Rowman & Littlefield, 1999.

National Center for Health Statistics. *Fastats: Cancer*. Hyattsville, MD: U.S. Government Printing Office, 2004. Also available at http://www.cdc.gov/nchs/fastats/cancer.htm (accessed 21 Feb. 2005).

———. *Fastats: Diabetes.* Hyattsville, MD: U.S. Government Printing Office, 2004. Also available at http://www.cdc.gov/nchs/fastats/diabetes.htm (accessed 21 Feb. 2005).

———. *Fastats: Heart Disease.* Hyattsville, MD: U.S. Government Printing Office, 2004. Also available at http://www.cdc.gov/nchs/fastats/heart.htm (accessed 21 Feb. 2005).

———. *Health, United States, 2004.* Hyattsville, MD: U.S. Government Printing Office, 2004. Also available at http://www.cdc.gov/nchs/data/hus/hus04.pdf (accessed 14 Jan. 2005).

Navaie-Waliser, M., A. Spriggs, and P. H. Feldman. "Informal Caregiving: Differential Experiences by Gender." *Medical Care* 40, no. 12 (2002): 1249–59.

Nelkin, D. "Behavioral Genetics and Dismantling the Welfare State." In *Behavioral Genetics: The Clash of Culture and Biology,* edited by R. Carson and M. Rothstein, 156–71. Baltimore, MD: Johns Hopkins University Press, 1999.

———. "Less Selfish Than Sacred? Genes and the Religious Impulse in Evolutionary Psychology." In *Alas, Poor Darwin: Arguments against Evolutionary Psychology,* edited by H. Rose and S. Rose, 17–32. New York: Harmony Books, 2000.

Nelkin, D., and S. Lindee. *The DNA Mystique: The Gene as a Cultural Icon.* New York: W. H. Freeman, 1985.

Nelkin, D., and L. Tancredi. *Dangerous Diagnostics: The Social Power of Biological Information.* Chicago: University of Chicago Press, 1995.

Nelson, J. L. "Prenatal Diagnosis, Personal Identity, and Disability." *Kennedy Institute of Ethics Journal* 10, no. 3 (2000): 213–28.

Nephew, K. P., and T. H. Huang. "Epigenetic Gene Silencing in Cancer Initiation and Progression." *Cancer Letters* 190, no. 2 (2003): 125–33.

Newman, J. E., et al. "Gender Differences in Psychosocial Reactions to Cystic Fibrosis Carrier Testing." *American Journal of Medical Genetics* 113, no. 2 (2002): 151–57.

New York Task Force on Life and the Law. *Assisted Reproductive Technologies: Analysis and Recommendations for Public Policy.* New York: The Task Force, April 1998.

Ng, R. K., and J. B. Gurdon. "Epigenetic Memory of Active Gene Transcription Is Inherited through Somatic Cell Nuclear Transfer." *Proceedings of the National Academy of Sciences USA* 102, no. 6 (2005): 1957–62.

Nsiah-Jefferson, L. "Reproductive Laws, Women of Color, and Low-Income Women." In *Reproductive Laws for the 1990s,* edited by S. Cohen and N. Taub, 23–67. Clifton, NJ: Humana Press, 1989.

Nussbaum, M. "Aristotle on Human Nature and the Foundations of Ethics." In *World, Mind, and Ethics: Essays on the Ethical Philosophy of Bernard Williams,* edited by J. E. J. Altham and R. Harrison, 86–131. Cambridge, UK: Cambridge University Press, 1995.

———. *Love's Knowledge.* Oxford, UK: Oxford University Press, 1990.

———. "Non-Relative Virtues: An Aristotelian Approach." In *The Quality of Life,* edited by M. Nussbaum and A. Sen, 1–6. New York: Oxford University Press, 1993.

O'Brien, M. *Politics of Reproduction.* London: Routledge Kegan and Paul, 1981.

Office of Technology Assessment. *Infertility, Medical and Social Choices.* Washington, DC: U.S. Government Printing Office, 1988.

Osada, H., and T. Takahashi. "Genetic Alterations of Multiple Tumor Suppressors and Oncogenes in the Carcinogenesis and Progression of Lung Cancer." *Oncogene* 21, no. 48 (2002): 7421–34.

Otlowski, M. F., S. D. Taylor, and K. K. Barlow-Stewart. "Major Study Commencing into Genetic Discrimination in Australia." *Journal of Law and Medicine* 10, no. 1 (2002): 41–48.

Overall, C. *Ethics and Human Reproduction.* Boston: Allen & Unwin, 1987.

Oyama, S. *Evolution's Eye: A Systems View of the Biology-Culture Divide.* Durham, NC: Duke University Press, 2000.

Parens, E., and A. Asch. *Prenatal Testing and Disability Rights.* Washington, DC: Georgetown University Press, 2000.

———. "The Disability Rights Critique of Prenatal Genetic Testing: Reflections and Recommendations." *Hastings Center Report* 29, no. 5 (1999): S1–22.

Parker, L. S. "Breast Cancer Genetic Screening and Critical Bioethics' Gaze." *Journal of Medicine and Philosophy* 20, no. 3 (1995): 313–37.

Parks, J. *No Place Like Home?* Bloomington, IN: Indiana University Press, 2003.

Parsons, E., and P. Atkinson. "Lay Constructions of Genetic Risk." *Sociology of Health and Illness* 14, no. 4 (1992): 437–55.

Paul, D. "The History of Newborn Phenylketonuria Screening in the U.S." App. 5 in *Promoting Safe and Effective Genetic Testing in the United States,* edited by N. A. Holtzman and M. S. Watson, 137–59. Washington, DC: NIH-DOE Working Group on the Ethical, Legal, and Social Implications of Human Genome Research, 1997.

———. *The Politics of Heredity: Essays on Eugenics, Biomedicine, and the Nature-Nurture Debate.* Albany, NY: State University of New York Press, 1998.

Peleg, L., et al. "Mutations of the Hexosaminidase A Gene in Ashkenazi and Non-Ashkenazi Jews." *Biochemical Medicine and Metabolic Biology* 52, no. 1 (1994): 22–26.

Pence, G. *Who's Afraid of Human Cloning?* Lanham, MD: Rowman & Littlefield, 1997.

Peters, T. *Playing God?* New York: Routledge, 1997.

Petersen, G. M., et al. "The Tay-Sachs Disease Gene in North American Jewish Populations: Geographic Variations and Origins." *American Journal of Human Genetics* 35, no. 6 (1983): 1258–69.

Peto, J. "Cancer Epidemiology in the Last Century and the Next Decade." *Nature* 411, no. 6835 (2001): 390–95.

Phoenix, A., A. Woollett, and E. Lloyd, eds. *Motherhood.* London: Sage, 1991.

Pinker, S. *The Blank Slate.* New York: Penguin, 2003.

Post, S. G., et al. "The Clinical Introduction of Genetic Testing for Alzheimer's Disease." *JAMA* 277, no. 10 (1997): 832–36.

President's Council on Bioethics. *Human Cloning and Human Dignity: An Ethical Inquiry.* Washington, DC: President's Council on Bioethics, 2002. Available at http://www.bioethics.gov/reports/cloningreport/pcbe_cloning_report.pdf (accessed 16 Jan. 2005).

Pullman, D. "Universalism, Particularism and the Ethics of Dignity." *Christian Bioethics* 7, no. 3 (2001): 333–58.

Purdy, L. "Genetics and Reproductive Risk: Can Having Children Be Immoral?" In *Reproducing Persons: Issues in Feminist Bioethics*, 39–49. Ithaca, NY: Cornell University Press, 1996.

——. "What Can Progress in Reproductive Technology Mean for Women?" *Journal of Medicine and Philosophy* 21, no. 5 (1996): 499–514.

Quinn, L. "Behavior and Biology." *The Journal of Cardiovascular Nursing* 18, no. 1 (2003): 62–68.

Raine, A. *The Psychopathology of Crime: Criminal Behavior as a Clinical Disorder*. New York: Academic Press, 1997.

Reik, W., and J. Walter. "Genomic Imprinting: Parental Influence on the Genome." *Nature Review Genetics* 2 (2001): 21–32.

Reindal, S. M. "Disability, Gene Therapy and Eugenics—a Challenge to John Harris." *Journal of Medical Ethics* 26, no. 2 (2000): 89–94.

Renard, J. P., et al. "Nuclear Transfer Technologies: Between Successes and Doubts." *Theriogenology* 57, no. 1 (2002): 203–22.

Rendtorff, J. D. "Basic Ethical Principles in European Bioethics and Biolaw: Autonomy, Dignity, Integrity and Vulnerability—Towards a Foundation of Bioethics and Biolaw." *Medicine, Health Care, and Philosophy* 5, no. 3 (2002): 235–44.

Rewers, M., and R. F. Hamman. "Risk Factors for Non-Insulin-Dependent Diabetes." In *Diabetes in America*, 179–220. Bethesda, MD: NIH, 1995.

Rhodes, R. "Genetic Links, Family Ties, and Social Bonds: Rights and Responsibilities in the Face of Genetic Knowledge." *Journal of Medicine and Philosophy* 23, no. 1 (1998): 10–30.

Richards, J. R. *Human Nature after Darwin: A Philosophical Introduction*. London: Routledge, 2000.

Richards, M. P. "Lay Understanding of Mendelian Genetics." *Endeavour* 22, no. 3 (1998): 93–94.

Richards, R. J. "A Defense of Evolutionary Ethics." *Biology and Philosophy* 1 (1986): 265–93.

Risch, N. "Molecular Epidemiology of Tay-Sachs Disease." *Advances in Genetics* 44 (2001): 233–52.

——. "Searching for Genetic Determinants in the New Millennium." *Nature* 405 (2002): 847–56.

Robert, J. S. "Interpreting the Homeobox: Metaphors of Gene Activation in Development and Evolution." *Evolution and Development* 3, no. 4 (2001): 287–95.

Robertson, J. "Human Cloning and the Challenge of Regulation." *The New England Journal of Medicine* 339, no. 2 (1998): 119–22.

——. "The Question of Human Cloning." *Hastings Center Report* 24, no. 2 (1994): 6–14.

Robins, R., and S. Metcalfe. "Integrating Genetics as Practices of Primary Care." *Social Science and Medicine* 59, no. 2 (2004): 223–33.

Roche, P., and G. Annas. "Protecting Genetic Privacy." *Nature Reviews: Genetics* 2, no. 5 (2001): 392–96.

Rose, H., and S. Rose. "Introduction." In *Alas, Poor Darwin: Arguments against Evolutionary Psychology*, edited by H. Rose and S. Rose, 1–16. New York: Harmony Books, 2000.

Rose, S. "Escaping Evolutionary Psychology." In *Alas, Poor Darwin: Arguments against Evolutionary Psychology*, edited by H. Rose and S. Rose, 299–320. New York: Harmony Books, 2000.

———. "Moving on from Old Dichotomies: Beyond Nature-Nurture towards a Lifeline Perspective." *British Journal of Psychiatry*, suppl. 40 (2001): S3–S7.

———. "The Poverty of Reductionism." In *Thinking about Evolution: Historical, Philosophical, and Political Perspectives*, vol. 2, edited by R. Singh et al., 415–28. Cambridge, UK: Cambridge University Press, 2001.

Rose, S., R. Lewontin, and L. Kamin. *Not in Our Genes: Biology, Ideology, and Human Nature*. New York: Pantheon, 1984.

Rosenberg, A. *Darwinism in Philosophy, Social Science, and Policy*. Cambridge, UK: Cambridge University Press, 2000.

Roses, A. D. "Apolipoprotein E Affects the Rate of Alzheimer's Disease Expression: Beta-Amyloid Burden Is a Secondary Consequence Dependent on APOE Genotype and Duration of Disease." *Journal of Neuropathology and Experimental Neurology* 53, no. 5 (1994): 429–37.

Ross, C. A., and R. L. Margolis. "Huntington's Disease." *Clinical Neuroscience Research* 1, nos. 1–2 (2001): 142–52.

Rothman, B. K. *Recreating Motherhood*. New York: W. W. Norton & Company, 1990.

———. *The Tentative Pregnancy*. New York: Viking, 1986.

Rothstein, M. A. "Behavioral Genetic Determinism: Its Effects on Culture and Law." In *Behavioral Genetics: The Clash of Culture and Biology*, edited by R. Carson and M. Rothstein, 89–115. Baltimore, MD: Johns Hopkins University Press, 1999.

Rothstein, M. A., and M. R. Anderlik. "What Is Genetic Discrimination, and When and How Can It Be Prevented?" *Genetics in Medicine* 3, no. 5 (2001): 354–58.

Rothstein, M. A., and S. Hoffman. "Genetic Testing, Genetic Medicine, and Managed Care." *Wake Forest Law Review* 34, no. 3 (1999): 849–88.

Rottschaefer, W. *The Biology and Psychology of Moral Agency*. Cambridge, UK: Cambridge University Press, 1998.

Rowland, R. *Living Laboratories*. Bloomington, IN: Indiana University Press, 1992.

Rozenberg, R., and V. Pereira. "The Frequency of Tay-Sachs Disease Causing Mutations in the Brazilian Jewish Population Justifies a Carrier Screening Program." *Sao Paulo Medical Journal* 119, no. 4 (2001): 146–49.

Ruse, M. "Reduction in Genetics." In *PSA 1974*, edited by R. S. Coen et al., 633–51. Dordrecht, the Netherlands: Reidel, 1976.

———. *Taking Darwin Seriously: A Naturalistic Approach to Philosophy*. Amherst, New York: Prometheus Books, 1998.

Saint-Paul, G. "Economic Aspects of Human Cloning and Reprogenetics." *Economic Policy* 18, no. 1 (2003): 73–122.

Sansinena, M. J., et al. "Production of Nuclear Transfer Llama (Lama Glama) Embryos from In Vitro Matured Llama Oocytes." *Cloning Stem Cells* 5, no. 3 (2003): 191–98.

Santos, F., and W. Dean. "Epigenetic Reprogramming during Early Development in Mammals." *Reproduction* 127, no. 6 (2004): 643–45.

Sarkar, S. *Genetics and Reductionism*. Cambridge, UK: Cambridge University Press, 1998.

Schachter, O. "Human Dignity as a Normative Concept." *American Journal of International Law* 77, no. 4 (1983): 848–54.

Schaffner, K. F. "Reductionism in Biology, Prospects and Problems." In *PSA 1974*, edited by R. S. Coen et al., 613–32. Dordrecht, the Netherlands: Reidel, 1976.

Schairer, C. F., et al. "Menopausal Estrogen and Estrogen-Progestin Replacement Therapy and Breast Cancer Risk." *JAMA* 283, no. 4 (2000): 485–91.

Schenker, J. G. "Clinical Aspects of Ovarian Hyperstimulation Syndrome." *European Journal of Obstetric, Gynecology and Reproductive Biology* 85, no. 1 (1999): 13–20.

Scriver, C. R. "Why Mutation Analysis Does Not Always Predict Clinical Consequences: Explanations in the Era of Genomics." *Journal of Pediatrics* 140, no. 5 (2002): 502–6.

Scriver, C. R., and P. J. Waters. "Monogenic Traits Are Not Simple: Lessons from Phenylketonuria." *Trends in Genetics* 15, no. 7 (1999): 267–72.

Senior, V., T. Marteau, and T. Peters. "Will Genetic Testing for Predisposition for Disease Result in Fatalism? A Qualitative Study of Parents' Responses to Neonatal Screening for Familial Hypercholesterolaemia." *Social Science and Medicine* 48, no. 12 (1999): 1857–60.

Senior, V., T. Marteau, and J. Weinman. "Impact of Genetic Testing on Causal Models of Heart Disease and Arthritis: An Analogue Study." *Psychology & Health* 14, no. 6 (2000): 1077–88.

Shakespeare, T., and M. Erickson. "Different Strokes: Beyond Biological Determinism and Social Constructivism." In *Alas, Poor Darwin: Arguments against Evolutionary Psychology*, edited by H. Rose and S. Rose, 229–48. New York: Harmony Books, 2000.

Sherwin, S. *No Longer Patient*. Philadelphia: Temple University Press, 1992.

Shiels, P. G., and A. G. Jardine, "Dolly, No Longer the Exception: Telomeres and Implications for Transplantation." *Cloning Stem Cells* 5, no. 2 (2003): 157–60.

Shushan, A., et al. "Human Menopausal Gonadotropin and the Risk of Epithelial Ovarian Cancer." *Fertility and Sterility* 65, no. 1 (1996): 13–18.

Shuster, E. "Human Cloning: Category, Dignity, and the Role of Bioethics." *Bioethics* 17, nos. 5–6 (2003): 517–25.

Silver, L. M. *Remaking Eden: Cloning and Beyond in a Brave New World*. New York: Avon, 1997.

Silvers, A., and M. A. Stein. "Human Rights and Genetic Discrimination: Protecting Genomics' Promise for Public Health." *Journal of Law, Medicine and Ethics* 31, no. 3 (2003): 377–89.

Silvers, A., D. Wasserman, and M. Mahowald. *Disability, Difference, Discrimination*. Lanham, MD: Rowman & Littlefield, 1998.

Singer, P. *A Darwinian Left: Politics, Evolution, and Cooperation*. New Haven, CT: Yale University Press, 2000.

Sober, E. *From a Biological Point of View*. Cambridge, UK: Cambridge University Press, 1984.

———. "The Meaning of Genetic Causation." In *From Chance to Choice: Genetics and Justice*, A. Buchanan, D. Brock, N. Daniels, and D. Wikler, 347–70. Cambridge, UK: Cambridge University Press, 2000.

Solter, D. "Mammalian Cloning: Advances and Limitations." *Nature Review: Genetics* 1, no. 3 (2000): 199–207.

Sommerville, A., and V. English. "Genetic Privacy: Orthodoxy or Oxymoron?" *Journal of Medical Ethics* 25, no. 2 (1999): 144–50.

Stacey, M. "The New Genetics: A Feminist View." In *The Troubled Helix: Social and Psychological Implications of the New Human Genetics*, edited by T. M. Marteau, and M. P. M. Richards, 331–49. Cambridge, UK: Cambridge University Press, 1996.

Steinbock, B., and R. McClamrock. "When Is Birth Unfair to the Child?" *The Hastings Center Report* 24, no. 6 (1994): 15–21.

Sterelny, K., and P. Kitcher. "The Return of the Gene." *The Journal of Philosophy* 85, no. 7 (1988): 339–61.

Stewart, A., and G. W. Kneale. "Radiation Dose Effects in Relation to Obstetrics, X Ray and Childhood Cancer." *Lancet* 1, no. 7658 (1970): 1185–87.

St. John, J. C., et al. "The Consequences of Nuclear Transfer for Mammalian Foetal Development and Offspring Survival: A Mitochondrial DNA Perspective." *Reproduction* 127, no. 6 (2004): 631–41.

St. John, J. C., R. Lloyd, and S. El Shourbagy. "The Potential Risks of Abnormal Transmission of MtDNA through Assisted Reproductive Technologies." *Reproductive Biomedicine Online* 8, no. 1 (2004): 34–44.

Strandell, A., J. Thorburn, and L. Hamberger. "Risk Factors for Ectopic Pregnancy in Assisted Reproduction." *Fertility and Sterility* 71, no. 2 (1999): 282–86.

Stuhrmann, M., et al. "Mutation Screening for Prenatal and Presymptomatic Diagnosis: Cystic Fibrosis and Haemochromatosis." *European Journal of Pediatrics* 159, suppl. 3 (2000): S186–S191.

Sutton, V. R. "Tay-Sachs Disease Screening and Counseling Families at Risk for Metabolic Disease." *Obstetrics and Gynecology Clinics of North America* 29, no. 2 (2002): 287–96.

Takala, T. "The Right to Genetic Ignorance Confirmed." *Bioethics* 13, nos. 3–4 (1999): 288–93.

Takala, T., and M. Häyry. "Genetic Ignorance, Moral Obligations and Social Duties." *Journal of Medicine and Philosophy* 25, no. 1 (2000): 107–13.

Tamada, H., and N. Kikyo. "Nuclear Reprogramming in Mammalian Somatic Cell Nuclear Cloning." *Cytogenetic and Genome Research* 105, nos. 2–4 (2004): 285–91.

Tauer, C. A. "Genetic Testing and Discrimination: How Can We Protect Job and Insurance Policy Applicants from Negative Test Consequence?" *Health Progress* 82, no. 2 (2001): 48–53, 71.

Taylor, P. J., and J. V. Kredentser. "Diagnostic and Therapeutic Laparoscopy and Hysteroscopy and Their Relationship to In Vitro Fertilization." In *A Textbook of In Vitro*

Fertilization and Assisted Reproductive Technology, edited by P. R. Brinsden and P. A. Rainsbury, 73–92. Park Ridge, NJ: Parthenon Publishing Group, 1992.

Thornhill, R., and C. T. Palmer. *A Natural History of Rape: Biological Basis of Sexual Coercion*. Cambridge, MA: The MIT Press, 2000.

Tiger, L. *Men in Groups*. New York: Vintage Books, 1970.

Tiger, L., and R. Fox. *The Imperial Animal*. New York: Holt, Rinehart, and Winston, 1971.

Tobin, S. L., et al. "The Genetics of Alzheimer Disease and the Application of Molecular Tests." *Genetic Testing* 3, no. 1 (1999): 37–45.

Toulmin, S. "The Tyranny of Principles." *Hastings Center Report* 11 (1981): 31–39.

Tuana, N. *The Less Noble Sex: Scientific, Religious, and Philosophical Conceptions of Woman's Nature*. Bloomington, IN: Indiana University Press, 1993.

UNESCO. *Universal Declaration on the Human Genome and Human Rights*. Paris, France: UNESCO, 1997. Also available at http://portal.unesco.org/en/ev.php-URL_ID=13177&URL_DO=DO_TOPIC&URL_SECTION=201.html (accessed 11 March 2005).

U.S. Congress, Committee on Energy and Commerce, Subcommittee on Commerce, Trade, and Consumer Protection. *The Potential for Discrimination in Health Insurance Based on Predictive Genetic Tests*. Washington, DC: U.S. Government Printing Office, 2002.

U.S. Congress, Senate, Committee on Health, Education, Labor, and Pensions. *Protecting Against Genetic Discrimination: The Limits of Existing Laws*. Washington, DC: U.S. Government Printing Office, 2002.

U.S. Department of Commerce, Economics and Statistics Administration, U.S. Census Bureau. *Health Insurance Coverage in the United States: 2002*. Washington, DC: U.S. Census Bureau, 2003. Also available at http://www.census.gov/prod/2003pubs/p60-223.pdf (accessed 5 Feb. 2005).

U.S. Department of Energy, Human Genome Project Information. *Gene Testing*. Available at http://www.ornl.gov/sci/techresources/Human_Genome/medicine/genetest.shtml (accessed 21 Feb. 2005).

U.S. Department of Energy, Office of Science, Human Genome Project Information. *About the Human Genome Project*. Available at http://www.ornl.gov/sci/tech resources/Human_Genome/project/about.shtml (accessed 16 Feb. 2005).

U.S. Department of Health and Human Services. *Understanding Genetic Testing*. Available at http://www.accessexcellence.org/AE/AEPC/NIH/ (accessed 21 Feb. 2005).

U.S. Department of Justice, Office of Justice Programs, Bureau of Justice Statistics. *Criminal Victimization 2003*, NCJ 205455. Washington, DC: Department of Justice, 2004. Also available at http://www.ojp.usdoj.gov/bjs/pub/pdf/cv03.pdf (accessed 31 Jan, 2005).

U.S. Department of Labor, Bureau of Labor Statistics, News, Bureau of Labor Statistics. *The Employment Situation: January 2005*. Washington, DC: BLS, 2005. Also available at http://www.bls.gov/news.release/pdf/empsit.pdf (accessed 10 March 2005).

U.S. National Bioethics Advisory Commission. *Cloning Human Beings: Report and Recommendations of the National Bioethics Advisory Commission.* Rockville, MD: The Commission, 1997.

U.S. National Library of Medicine, National Institutes of Health. *Genetics Home Reference.* Bethesda, MD: 2004. Also available at http://ghr.nlm.nih.gov/info=genetic_testing/show/cost_results;jsessionid=4F0A6C9A1FEC9D2B52043E37E4792F48 (accessed 21 Feb. 2005).

Vehmas, S. "Just Ignore It? Parents and Genetic Information." *Theoretical Medicine* 22, no. 5 (2001): 473–84.

———. "Parental Responsibility and the Morality of Selective Abortion." *Ethical Theory and Moral Practice* 5, no. 4 (2002): 463–84.

Venn, A. "Risk of Cancer after Use of Fertility Drugs with In-Vitro Fertilization." *Lancet* 354, no. 9190 (1999): 1586–90.

Verhey, A. D. "Cloning: Revisiting an Old Debate." *Kennedy Institute of Ethics Journal* 4 (1994): 227–34.

Vineis, P., et al. "Misconceptions about the Use of Genetic Tests in Populations." *Lancet* 357, no. 9257 (2001): 709–12.

Vogel, G. "Human Cloning: Scientists Take Step toward Therapeutic Cloning." *Science* 303, no. 5660 (2004): 937–39.

Vuckovic, N., et al. "Consumer Knowledge and Opinions of Genetic Testing for Breast Cancer Risk." *American Journal of Obstetrics and Gynecology* 189, no. 4 (2003): S48–S53.

Wachbroit, R. "The Question Not Asked: The Challenge of Pleiotropic Genetic Tests." *Kennedy Institute of Ethics Journal* 8, no. 2 (1998): 131–44.

Walker, J. S. "The Controversy over Radiation Safety: A Historical Overview." *JAMA* 262, no. 5 (1989): 664–68.

Walker, M. *Moral Understandings: A Feminist Study in Ethics.* New York: Routledge, 1998.

Wall, T. L., et al. "Alcohol Dehydrogenase Polymorphisms in Native Americans: Identification of the ADH2*3 Allele." *Alcohol and Alcoholism* 32, no. 2 (1997): 129–32.

Wall, T. L., L. G. Carr, and C. L. Ehlers. "Protective Association of Genetic Variation in Alcohol Dehydrogenase with Alcohol Dependence in Native American Mission Indians." *The American Journal of Psychiatry* 160, no. 1 (2003): 41–46.

Wallace, D. C. "Mitochondrial DNA Mutations in Diseases of Energy Metabolism." *Journal of Bioenergetics and Biomembranes* 26, no. 3 (1994): 241–50.

Walters, L., and J. G. Palmer. *The Ethics of Human Gene Therapy.* New York: Oxford University Press, 1997.

Warnock, M. *A Question of Life: The Warnock Report on Human Fertilization and Embryology.* Oxford, UK: Blackwell, 1985.

Waters, C. K. "Genes Made Molecular." *Philosophy of Science* 61, no. 2 (1994): 163–85.

———. "Why the Anti-Reductionist Consensus Won't Survive: The Case of Classical Mendelian Genetics." In *PSA 1990*, edited by A. Fine, M. Forbes, and L. Wessels, 125–39. East Lansing, MI: Philosophy of Science Association, 1990.

Watson, E. K., et al. "Psychological and Social Consequences of Community Carrier Testing Screening for Cystic Fibrosis." *Lancet* 340 (1992): 217–20.

Watson, J. *A Passion for DNA: Genes, Genome, and Society.* New York: CSHL Press, 2000.

Watson, J., and F. Crick. "A Structure for Deoxyribose Nucleic Acid." *Nature* 171 (1953): 737–38.

Watson, J., et al. *Recombinant DNA.* 2nd ed. New York: W. H. Freeman, 1992.

Weinstein, N. D., and W. M. Klein. "Resistance of Personal Risk Perceptions to Debiasing Interventions." *Health Psychology* 14, no. 2 (1995): 132–40.

Weiss, K., and A. Buchanan. "Evolution by Phenotype: A Biomedical Perspective." *Perspectives in Biology & Medicine* 46, no. 2 (2003): 159–82.

Weissenbach, J. "The Human Genome Project: From Mapping to Sequencing." *Clinical Chemistry and Laboratory Medicine* 36, no. 8 (1998): 511–14.

Weitzel, N. "The Current Social, Political, and Medical Role of Genetic Testing in Familial Breast and Ovarian Carcinomas." *Current Opinion in Obstetrics and Gynecology* 11, no. 1 (1999): 65–70.

Welch, H., and W. Burke. "Uncertainties in Genetic Testing for Chronic Disease." *JAMA* 280, no. 17 (1998): 1525–27.

Welcsh, P. L., and M. C. King. "BRCA1 and BRCA2 and the Genetics of Breast and Ovarian Cancer." *Human Molecular Genetics* 10, no. 7 (2001): 705–13.

Wellington, C. L., et al. "Toward Understanding the Molecular Pathology of Huntington's Disease." *Brain Pathology* 7, no. 3 (1997): 979–1002.

Wells, D., and D. A. Delhanty. "Preimplantation Genetic Diagnosis: Applications for Molecular Medicine." *Trends in Molecular Medicine* 7, no. 1 (2001): 23–30.

Wexler, A. *Mapping Fate.* Berkeley, CA: University of California Press, 1996.

Wilkie, A. "Genetic Prediction: What Are the Limits?" *Studies in the History and Philosophy of Biological and Biomedical Sciences* 32, no. 4 (2001): 619–33.

Wilmut, I., and L. Paterson. "Somatic Cell Nuclear Transfer." *Oncology Research* 13, nos. 6–10 (2003): 303–07.

Wilmut, I., K. Campbell, and C. Tudge. *The Second Creation.* Cambridge, MA: Harvard University Press, 2000.

Wilmut, I., et al. "Viable Offspring Derived from Fetal and Adult Mammalian Cells." *Nature* 385 (1997): 810–13.

Wilson, E. O. *On Human Nature.* Cambridge, MA: Harvard University Press, 1978.

Wing, R. R., et al. "Behavioral Science Research in Diabetes." *Diabetes Care* 24, no. 1 (2001): 117–23.

Winston, R. "The Promise of Cloning for Human Medicine." *British Medical Journal* 314, no. 7085 (1997): 913–14.

Wolf, S., ed. *Feminism and Bioethics.* New York: Oxford University Press, 1996.

Woolcock, P. "The Case Against Evolutionary Ethics Today." In *Biology and the Foundation of Ethics*, edited by J. Maienschein and M. Ruse, 276–306. Cambridge, UK: Cambridge University Press, 1999.

World Health Organization. *Ethical, Scientific and Social Implications of Cloning in Human Health.* Geneva, Switzerland: WHO, 1998.

Wrangham, R., and D. Peterson. *Demonic Males: Apes and the Origins of Violence.* Boston: Houghton Mifflin, 1996.

Wright, R. *The Moral Animal: Why We Are the Way We Are; The New Science of Evolutionary Psychology.* New York: Pantheon, 1994.

Wright, S. "Recombinant DNA Technology and Its Social Transformation 1972–1982." *Osiris* 2 (1986): 303–60.

Wuest, J. "Repatterning Care: Women's Proactive Management of Family Caregiving Demands." *Health Care for Women International* 21, no. 5 (2000): 393–411.

Yamanaka, S., et al. "Structure and Expression of the Mouse Beta-Hexosaminidase Genes, *Hexa* and *Hexb*." *Genomics* 21, no. 3 (1994): 588–96.

Yanagimachi, R. "Cloning: Experience from the Mouse and Other Animals." *Molecular and Cellular Endocrinology* 187, nos. 1–2 (2002): 241–48.

Zinberg, R. E., et al. "Prenatal Genetic Screening in the Ashkenazi Jewish Population." *Clinical Perinatology* 28, no. 2 (2001): 367–82.

Index

About the Author

Inmaculada de Melo-Martín, previously Associate Professor of Philosophy at St. Mary's University in San Antonio, Texas, is now a Research Ethicist in the Division of Medical Ethics in the Department of Public Health, Weill Medical College of Cornell University. She is the author of *Making Babies* (Dordrecht: Kluwer, 1998). Dr. de Melo-Martín's work has appeared in *Bioethics, Philosophy of Science,* and *Studies in the History and Philosophy of Biology and the Biomedical Sciences.*